Modelling and Measuring Reactor Core Graphite Properties and Performance

Modelling and Measuring Reactor Core Graphite Properties and Performance

Edited by

Gareth B. Neighbour
Department of Mechanical Engineering and Mathematical Sciences,
Oxford Brookes University, Oxford, UK
Email: gneighbour@brookes.ac.uk

RSC Publishing

The proceedings of the meeting on Modelling and Measuring Reactor Core Graphite Properties and performance held 31st October – 3rd November 2011 at Aston University, United Kingdom, and run under the auspices of the British Carbon Group.

Special Publication No. 342

ISBN: 978-1-84973-390-8

A catalogue record for this book is available from the British Library

Published by The Royal Society of Chemistry,
Thomas Graham House, Science Park, Milton Road,
Cambridge CB4 0WF, UK

Registered Charity Number 207890

Visit our website at www.rsc.org/books

Printed in the United Kingdom by CPI Group (UK) Ltd, Croydon, CR0 4YY, UK

Foreword

The United Kingdom has significant experience in the operation of civil graphite-moderated nuclear reactors for the production of electricity. The experience is certainly unique in terms of the problems faced in normal operation and the challenges raised through ageing mechanisms. It is important that we recognize the significant acheivements todate. In particular, for AGRs, the concern is with brick cracking (and subsequent change in core behaviour) and graphite weight loss. Consequently, the safety requirements for the AGRs can be expressed as: unhindered movement of control rods; adequate cooling of fuel and core; and, unhindered fuel charge and discharge.

The achievement of the utility companies and their supporting organisations in maintaining the integrity and functionality of the core is something that cannot have been predicted in 1956 when the first Calder Hall connected to the national grid. Whilst the final curtain approaches for the fleet of Magnox nuclear reactors, the fleet of Advanced Gas-cooled Reactors, where the first reactors entered service in 1976, approach the most thought-provoking juncture of their lives where aspects of plant life extension are examined to demonstrate safe operation and enhanced economic performance possibly until 2030 and of course this will also mitigate against the generation of CO_2. There are significant pieces of novel science and engineering work on-going to underpin safety cases and this work is captured and reflected in this book. As the UK, indeed the world, contemplates the next (3rd) generation of nuclear power and the likely move away from graphite-moderated reactors in favour of pressurized water reactors, graphite moderated reactors will undoubtedly return in Generation 4 as the need to have higher temperatures to increase performance and produce process steam such as that used in the production of hydrogen. This book will no doubt be useful as designers today contemplate the issues of future graphite reactors, especially the thoughts captured in the discussion which serve as a warning to future designers.

Gareth B. Neighbour
Oxford Brookes University

Preface

The United Kingdom has a number of gas-cooled, graphite-moderated nuclear reactors that are either nearing the end of their generating life, seeking life extension, or in the latter period of their design lives. All graphite moderated reactors suffer ageing and degradation to the graphite during service. The degradation poses a threat to the functionality of the graphite core, and potentially, the safe operation of the reactor. As the graphite components of the reactor cores cannot be replaced, it is important to ensure that effective strategies are in place to secure safe and reliable operation ultimately beyond their planned design life. The importance of extending the safe operating life of a reactor is increasing due to the political and socio-economic demands to reduce greenhouses gases and to diversify energy supply. Thus, the importance of modelling and measuring reactor core graphite properties and performance increases as life extension becomes significant.

This book represents the record and output from the 'Modelling and Measuring Reactor Core Graphite Properties and Performance' conference which took place between the 31st October and 3rd November 2011 at Aston University. The conference was very well attended and was organised with support from all the major stakeholders in the UK, namely AMEC, Atkins, EDF Energy, Frazer Nash Consultants, Health and Sagety Executive, Magnox, National Nuclear Laboratory, NRG, and Serco. This conference represented a third gathering of the most eminent specialists in the field.

The objective of the present book is to display the extensive research and analysis performed in support of the UK Advanced Gas-Cooled Reactors in regard to the integrity and continuing safe performance of the graphite cores. A key feature of this book are the reports on the debates that took place during the conference so that the challenges within each topic area discussed and collectively the scientific community can identify where there are deficiencies in the coolective understanding. Another way to consider this is the range of tools and tecniques employed in achieveing plant life extension have developed understanding form the atomic scale to that for full-scale compents. The question arises whether the operational life and thus the criteria for end-of-life can be fully expressed as the understanding of the inter-relationships of established mechansims, statistical inferences, empirical models and plant performance is extended.

It is also important to recognise that this book (which follows the conference at Aston University) is the third in a series with the first two being based on conferences held in Cardiff and Nottingham in 2005 and 2008, respectively. The proceedings of these conferences can be found in the following books.

1. Neighbour, G. B. (Ed) (2007). *Management of Ageing in Graphite Reactor Cores* (308 pp). Royal Society of Chemistry, Cambridge, (ISBN-10: 0854043454; ISBN-13: 978-0854043453).
2. Neighbour, G. B. (Ed) (2010). *Securing the Safe Performance of Graphite Reactor Cores* (270 pp). Royal Society of Chemistry, Cambridge, (ISBN: 978-1-84755-913-5).

The current book is divided into four topic areas related to modelling or measurement of the reactor core: Mechanistic; Statistical; Empirical; and, Plant. At the end of the book, there is a record of the open discussion that took place at the conference and reflects each topic area. The insights brought out of the discussion are very valuable and serve as lessons for future reactor designers.

Gareth B. Neighbour
Oxford Brookes University

Acknowledgements

The editor would like to express his heartfelt thanks to the many authors who contributed to this volume. He would also like to thank his wife Angela and children, Caitlin and Rhiannon, for their love, support and tolerance in preparing this book. Finally, but not least, the editor would like to recognise the support of the British Carbon Group (registered charity 207890 which itself is affiliated to The Royal Society of Chemistry, The Institute of Physics and The Society of Chemical Industry) and the Organising Committee of the conference without which this book would not exist.

Organising Committee

Prof Peter Flewitt	*University of Bristol & Conference Chair*
Dr James Reed	*EDF Energy*
Dr Gareth Neighbour (Editor)	*Oxford Brookes University & Chairman, British Carbon Group*
Dr Anthony Wickham (Secretary)	*Conference Manager*

On behalf of the British Carbon Group, the editor would also like to express thanks to the following organisation for financial support for the conference which led to this book.

- EDF Energy
- AMEC
- Atkins
- Frazer Nash Consultants
- Health and Safety Executive
- Magnox Ltd
- National Nuclear Laboratory
- NRG
- Serco

Contents

Part A – Mechanistic

Part B - Empirical

Part A

Mechanistic

Towards a Structural Basis to the Physical Properties of Irradiated Polycrystalline Nuclear Graphite

Brian Rand
Institute for Materials Research, University of Leeds
and
Institute of Applied Materials (SARChI Chair of Carbon Technology and Materials), University of Pretoria, RSA
Email: b.ca.rand@btinternet.com

Abstract
The need for mechanistic understanding of structure-property relationships for irradiated polycrystalline nuclear graphite is considered. Appropriate structural characteristics for polycrystalline graphite at all of the relevant dimensional scales are discussed and some aspects of the effects of irradiation damage on them are considered. It is suggested that the relationship between structural characteristics and physical properties be determined at each dimensional scale and used in multi-scale modelling of specific properties at the macro-scale at which they are actually measured.

Keywords
Graphite, structure, properties

INTRODUCTION

Although significant progress was made in the periods when graphite core reactors were being developed and constructed, the following period has until recently been one of limited research activity due to the political perception of nuclear safety issues. Yet this very period is one when advanced material characterisation techniques have made tremendous strides along with the development of structural modelling techniques. Now that there is renewed interest in developing an improved understanding of the structure-property relationships for polycrystalline graphite, some of the earlier concepts inevitably will be re-evaluated. Currently, many of the material models in use to predict the changes in physical properties of nuclear graphite resulting from fast neutron irradiation and radiolytic oxidation are empirically based and their implementation is limited to the range of fluence and oxidative weight loss for which there is a validated database, hence the need for better mechanistic understanding of the structural basis of the properties and the development of mechanistically based models for the different properties. This paper presents a personal view of what may be required to develop such an approach.

STRUCTURE-PROPERTY RELATIONSHIPS

For the UK situation, perhaps the most important mechanistic issues for polycrystalline material are:-

- The mechanism of dimensional change, since the differential changes control internal stresses;
- Irradiation creep which leads to the alleviation of such stresses;
- The structural changes due to fast neutron irradiation and to radiolytic oxidation;
- Structural dependence of physical properties.

Given this level of understanding, it should be possible to understand the effects of irradiation damage and oxidation on the physical properties of interest. Such a task, however, is not easy due to the structural complexity of polycrystalline graphite. Figure 1 shows the obvious relationship between structural characteristics and physical properties. When the former are changed by irradiation and oxidation, the physical properties are changed. However, if we cannot clearly define what these structural characteristics are and/or do not know how they are changed by irradiation then the modified physical properties cannot be predicted. Consequently, because we do not currently have this level of understanding and because it has been relatively easy to determine the changes in physical properties the tendency has been to use the physical property database developed for prediction of core component behaviour and then for mechanistic understanding to attempt to deduce how the structure has been modified, as shown by the grey arrows in Figure 1.

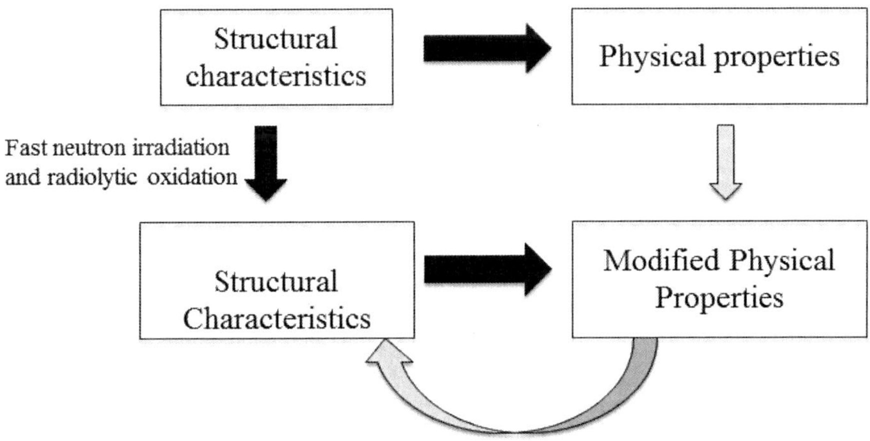

FIGURE 1: Dependence of physical properties on structural characteristics.

Such an approach is inevitably limited in its scope. The essential database of properties is created, but its ability to be extrapolated to fluences much beyond the extent of the database is severely limited. The level of understanding has also been supplemented by the study of materials approximating to the single crystal state and whilst such studies have invaluable, they also are limited in providing insight only into the behaviour of the material at the crystallographic level as explained in more detail below.

STRUCTURE OF POLYCRYSTALLINE GRAPHITE

It is important first of all to define the important structural characteristics that will determine the relevant physical properties. This is not straight forward as there are important structural features at all dimensional scales, ranging from nano- to micro- to macroscopic that may in principle contribute to a particular physical property, the relative importance of each being different for each property. It is relatively easy to describe these qualitatively as below but more problematic to define quantitative characteristics that can be measured and related directly to specific physic-chemical properties, yet such an approach is required.

- At the nano-scale there are the regions showing coherent x-ray diffraction, often known as 'crystallites', defined by the coherence lengths in the two principal crystallographic directions, L_a and L_c and the interlayer spacing, d_{002}. There may also be nano-scale fissures associated with (separating) such regions arising perhaps from thermal contraction stresses during cool down from graphitization temperatures. The behaviour of this fundamental unit is usually understood from studies of natural flake graphites and highly oriented pyrolytic graphites (HOPG) (Kelly,1981). A range of basal and non-basal dislocations has been identified and the effects of irradiation damage at typical reactor temperatures in creating dislocation loops was studied in some detail (Kelly, 1981; Thrower, 1969). The dimensional changes accompanying irradiation have been documented at various temperatures (Kelly, 1981). However, the existence of such dislocations in nuclear graphites has not been well documented, there being limited recorded observations of dislocation for such material (Shtrombakh *et al.*, 1995). Their role is also not so clear due to the more limited extent of the coherent regions and the uncertain nature of the boundaries between them. Modern advanced characterisation techniques should in future enable a better understanding of the defect/dislocation character of irradiated nuclear graphites. Quantitative characteristics might be:

- L_a from x-ray diffraction and Raman spectroscopy.
- L_c and d_{002} from x-ray diffraction.
- Dislocation loop density.
- sp^2 character.

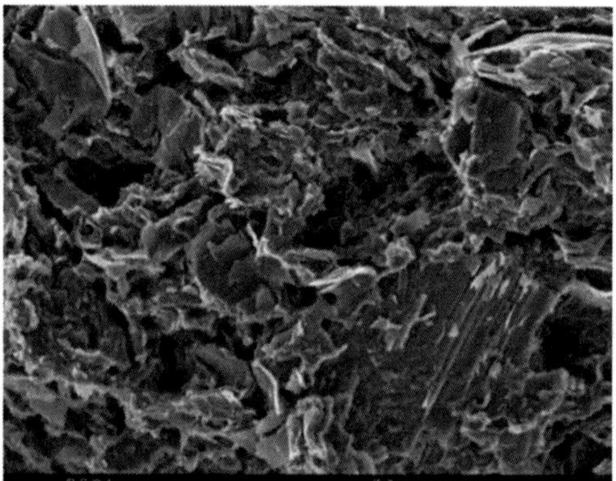

FIGURE 2: SEM micrograph of a fracture surface of PGA graphite showing domains of preferred orientation, cracks and pores.

- These coherence regions are not isolated but are connected to form domains of microscopic dimensions, extending over tens to hundreds of microns and visible in the optical microscope under crossed polars. Figure 2 shows a scanning electron micrograph (SEM) of such a region. A highly ordered 'needle' particle is shown in the bottom right hand corner and there are twisted layer packets or domains in the remainder of the view. In some publications, these domains (or regions within the domains, Shtrombakh *et al.* (1995)) have been taken to represent the crystallites and there is some justification for this as discussed below. The domains may be bent and twisted and within them there may be additional lamellar fissures, again arising from thermal contraction stresses, but also from the shrinkage arising from the densification of the graphite during manufacture as the true density increases from about 1.4 at the mesophase state where the lamelliform structure is developed through about 1.9 on 'baking' to 2.25 g cm^{-3} at the graphite stage. These fissures are understood to affect the thermal expansion behaviour and the dimensional change by partly accommodating the c-axis expansion and the c-axis dimensional change. Those fissures arising from thermal contraction (the so-called 'Mrozowski' cracks) may close progressively on heating whilst those arising from structural densification may not. Both, however, will contribute to the dimensional change. Thus, there is a local (micro-scale) preferred orientation of the coherence regions and the associated fissures that needs to be understood. The fundamental understanding that we have for the behaviour of 'crystallites' (derived from the study of HOPG for example) may more precisely relate to regions of connected coherence lengths with a preferred orientation lying within the larger domains and between the lamellar fissures and this will be the basic driving force for dimensional change. The behaviour of the domain, however, will be different due to the moderating effect of the fissures on expansion behaviour. The physical properties locally within a domain will also be different from those locally within the coherence region and of the connected coherence region. Quantitative parameters include:

- Domain dimensions in different directions (shape).
- Preferred orientation of layer planes within the domains (extent of wrinkling).
- Intra-domain porosity (fissures).
- Orientation of porosity.
- Local physical properties.

There will be a connection between the structural parameters at the nano-scale and some of the domain parameters. For example, the local preferred orientation within a domain will be changed as the coherence lengths are reduced by irradiation damage.

- At the macro-scale, there is the arrangement of domains. At some dimensional scale, there will be a preferred orientation of the domains, perhaps within the boundaries of a filler grain, but on a larger scale of size the orientation of domains may be isotropic or there may be a preferred orientation as in PGA graphite. At the macro-scale, we have to consider the filler and binder regions, which will show different domain textures although not necessarily vastly different crystallographic structure (*i.e.,* in terms of the coherently diffracting regions). It is certainly not true that the binder regions are amorphous as has occasionally been claimed. Quantitative parameters could comprise:

    - Filler grain/binder proportions.
    - Filler grain size/shape distribution.
    - Preferred orientation of anisometric filler grains.
    - Preferred orientation of domains within filler grains.
    - Domain dimensions within filler and binder.
    - Intra-porosity of filler grains.
    - Inter filler porosity.

So, whilst much has been made of the changes at the so-called 'crystallite' level, which may refer to coherence lengths or to domains, and it is surely the case that the driving force for structural rearrangement derives at this scale, in determining the effects on a macroscopic property some model characteristic structural unit at the macroscopic scale, which is a composite of the above structural features, will be the key to modeling a particular physical property.

EFFECT OF IRRADIATION ON STRUCTURE

The standard model of the effects of fast neutron irradiation on the structure (see, for example, Kelly (1981)) considers that interstitial carbon atoms form interplanar discs (dislocation loops) which lead to expansion of the crystal in the c-direction whilst vacancies created condense into lines leading to contraction in the a-direction. However, this view is currently questioned by Heggie and co-workers (2011), who have proposed a so-called 'ruck and tuck' model of damage and dimensional change operative at temperatures typical of reactor operation. The model involves the shearing of basal planes by dislocation movement. The validity of this new approach remains to be established although it has many attractive features. X-ray and transmission electron microscopy studies (TEM) have established that the coherent lengths show a drastic reduction after exposure to significant fluences with the TEM images showing clear 'wrinkling' of the layer planes (Asthana *et al.*, 2005) which resembles the structures in the coke at the carbon stage prior to complete graphitization of the material, *i.e.* the material becomes almost turbostratic (Tanabe and Muto, 1999). This raises the questions, "to what extent can irradiation damage be regarded as a reverse of the graphitization process" and "how valid is the 'ruck and tuck' model to graphitization itself"? The break up of the crystallographic structure creates many more boundaries between the coherently diffracting regions and, within the domains discussed above, there should be a corresponding change in the local preferred orientation. This structural change to a turbostratic state and the creation of these boundaries raises the question of the extent to which dislocation movement is relevant to the behaviour of irradiated graphite. A current model of irradiation creep by Kelly and Foreman (1974), for example, proposes that the secondary creep process is controlled by the pinning and unpinning of dislocations by neutron bombardment, but their paper acknowledges that *"A dislocation cannot be defined in a turbo-stratic structure."* Thus, the role of dislocations in determining the properties of graphite irradiated to high fluence requires clarification.

It is commonly accepted that there are changes to the network of fissures and more equi-axed pores that comprise the porosity in the material. This is well documented by the study of Engle (1971) who used a combination of pycnometric techniques to quantitatively assess the changes to various types of pores in two graphites through the region at which dimensional change turnaround occurs and beyond. He was able to show how pores between crystallite clusters were closed by the c-axis expansion and other pores opened up between crystallite clusters. Beyond turnaround in the macroscopic expansion phase, new pores were also formed between filler particles. However, the body of evidence for the structural changes through the whole extent of fluence likely to be experienced during reactor service is sparse and needs to be improved with the application of the modern methods available these days.

MODELLING OF GRAPHITE MATERIALS PROPERTIES

The new materials modeling approach adopted by EDF Energy (Bradford and Steer, 2008) attempts to take a unified approach to the physical properties of graphite, recognizing that porosity may be both eliminated and created, but it still has terms which have a poorly defined structural basis, for example the underlying concept of connectivity. On the other hand, it does make use of the known behaviour of HOPG crystals as driving forces for dimensional change. What is needed ideally is for this information from HOPG studies to feed into the model of domains and their interaction. Berre *et al.* (2008) applied Eshelby's micromechanics theory and the mean-field method to the prediction of the effect of elastic properties of filler particles and inter-granular porosity on the Young's modulus of graphite. Wang and Yu (2008) also considered the effect of porosity with different pore aspect ratios on the thermal and elastic properties of graphite using a large-scale three-dimensional boundary element model. Berre, Fok *et al.* (2008) applied multi-phase microstructural modelling to x-ray tomography images to account for the effects of porosity on mechanical properties, whilst Hall *et al.* (2006) created idealized finite element models of graphite filler particles and polycrystalline structures to model the changes in properties with fast neutron fluence. Panyukov and Subbotin (2008) have developed an analytical theory which relates the development of internal stresses with the evolution of an ensemble of micro-cracks and is based on an analysis of an orthotropic continuum with randomly arranged shape-changing crystallites. They were able to account for the observed dimensional changes and elastic modulus of a graphite as a function of fluence.

With the rapid development in integrated computational materials engineering (integrating models at a range of length scales), it should be possible to define a characteristic macroscopic structural unit that incorporates the essential micro- and nano-structural features defined above and which can be used to develop structure-property relationships for each property of interest. The scale for this modeling unit must incorporate all the features defined above and thus be a characteristic volume of material that incorporates both filler and binder regions. The scale of this characteristic unit will depend upon the filler grain size of the graphite. The structural changes at the macro level become increasingly important at medium to high fluences approaching and exceeding the dimensional change 'turnaround' region as it is here that it is considered that significant changes to the porosity are taking place whereas at low fluences, it may be that the dominant structural changes are at the domain level with only limited modification of the macro structure. However, it is precisely in the region of high fluences that data is lacking at the present time as much of the scientific study has concentrated on the behaviour of crystallites as modelled by that of HOPG.

Structure-property models at each relevant scale of size will be required, but modern structural characterization techniques are available to provide such information. The task is, however, large and if it is to be achieved will require a co-coordinated multi-disciplined approach. Some progress can be made with small groups perhaps concentrating on specific length scales or techniques and developing the local structure property relationships.

References

Asthana, A., Matsui, Y. T., Yasuda, M., Kimoto, K., Iwata, T. and Ohshima, K. (2005). Investigation on the structural disordering of neutron irradiated highly oriented pyrolytic graphite by x-ray diffraction and electron microscopy. *J. Applied Crystallography,* 38, 361-367.

Berre, C., Fok, S. L., Marsden, B. J., Mummery, P. M., Marrow, T. J. and Neighbour, G. B. (2008). Microstructural modelling of nuclear graphite using multi-phase models. *J. Nuclear Materials,* 380, 46-58.

Berre, C., Mummery, P. M., Marsden, B. J., Mori, T. and Withers, P. J. (2008). Application of a micromechanics model to the overall properties of heterogeneous graphite. *J. Nuclear Materials,* 381, 124-128.

Bradford, M. R. and Steer, A. G. (2008). A structurally based model of irradiated graphite properties. *J. Nuclear Materials,* 381, 39-45.

Engle, G. B. (1971). Irradiation behaviour of nuclear graphites at elevated temperatures. *Carbon,* 9, 539-554.

Hall, G., Marsden, B. J. and Fok, S. L. (2006). The microstructural modelling of nuclear grade graphite. *J. Nuclear Materials,* 353, 12-18.

Heggie, M. I., Suarez-Martinez, I., Davidson, C. and Haffenden, G. (2011). Buckle, ruck and tuck: A proposed new model for the response of graphite to neutron irradiation. *J. Nuclear Materials,* 413, 150-155.

Kelly, B. T. and Foreman, A. J. E. (1974). The theory of irradiation creep in reactor graphite – The dislocation pinning-unpinning model. *Carbon,* 12, 151-158.

Kelly, B. T. (1981). Physics of Graphite. Applied Science Publishers Ltd, Barking, UK. ISBN 0-85334-960-6.

Panyukov, S. V. and Subbotin, A. V. (2008). Theory of radiation induced shape change of graphite. *Atomic Energy,* 105, 32-41.

Shtrombakh, Ya. I., Gurovich, B. A., Platonov, P. A. and Alekseev, V. M. (1995). Radiation damage of graphite and carbon-graphite materials. *J. Nuclear Materials,* 225, 273-301.

Tanabe, T. and Muto, S. (1999). Damaging process of graphite – A new model and its impact on degradation of materials performance. *Physica Scripta,* T81, 104-107.

Thrower, P. A. (1969). The study of defects in graphite by Transmission Electron Microscopy. *Chemistry and Physics of Carbon,* 5, 217-319.

Wang, H. T. and Yu, S. Y. (2008). Large scale numerical simulation of mechanical and thermal properties of nuclear graphite using a microstructure-based model. *Nuclear Engineering and Design,* 238, 3203-3207.

Property Changes of Polycrystalline Graphite due to Neutron Irradiation – A Critical Assessment after 70 years of Research

Gerd Haag
Nuclear Technology Consultancy, Am Malefinkbach 3, D-52441 Linnich (Germany)
Email: gdhaag@t-online.de

Abstract
Research on the behaviour of polycrystalline graphite exposed to neutrons from nuclear fission reactions began around the year 1942. Up to today, many serious questions concerning properties and structural changes have been answered and in many countries graphite became a standard material for the core of gas-cooled reactors. However, the basis for that is a broad knowledge about manufacturing and a huge amount of numerical data from irradiation testing. After many years of successful use of graphite, new problems concerning the lifetime of graphite parts have arisen. In order to deal with these challenges a critical review is necessary of what we know, and of what is only subject to speculation. Difficulties with the interpretation of experimental data using classical models are discussed, and some new speculative ideas are presented.

Keywords
Nuclear graphite, irradiation behaviour, structural changes

INTRODUCTION

The use of graphite as a nuclear material was initiated by the first nuclear reactor project Chicago Pile 1 (CP1) reaching its first criticality under the leadership of Enrico Fermi in December 1942. At that time, the decision to choose graphite as the moderator was very easy as graphite in the form of electrode graphite was the only suitable material available. Purity was the problem to start with. First attempts to determine the neutron multiplication factor k gave values of about 0.87 showing that values of $k > 1$ could be made using purer graphite and purer uranium fuel. Therefore, purity was the first property in the focus of material development. Two measures were known to decrease the content of relevant impurities of graphite. Normally, electrode graphite is heat-treated to 2500-2600°C; graphitisation temperatures of 2800°C or higher would remove more impurities introduced by the raw materials which on their part had to be chosen carefully.

Already in December 1942, E. P. Wigner called attention to the effect of intense neutron radiation on pile material properties ("Wigner effect"). Early calculations indicated that one 2 MeV neutron produced in fission would possess enough energy to displace about 2000 carbon atoms. The possible effects on various material properties such as dimensions, strength, thermal conductivity, *etc.* are of much concern up to today. Despite the fact that quantum mechanics applied to crystalline solids has become a powerful tool, it remained largely futile from a practical standpoint. Predictions of what might happen to graphite due to fast neutron irradiation could only be made empirically, but the only neutron sources available were cyclotrons, and irradiations of several months were necessary to create effects equivalent to one day operation of the power reactors already under construction. The first really meaningful graphite irradiation experiments were performed in the Oak Ridge X-10 reactor finding out after six months' operation that the Hanford reactors could be operated for at least a week.

Many graphite reactors have been built since the early days. Large changes of many graphite properties have been observed and accounted for in subsequent reactor projects, whilst much effort has been invested to understand effects and to develop and manufacture better graphites. Nevertheless, it may be useful to ask ourselves today what we know for sure, what we believe we know, and what is still subject to speculation. Some stimulation shall be given in this paper.

DIMENSIONAL CHANGES OF GRAPHITE CRYSTALLITES

According to the dominant opinion of many graphite specialists physical properties of poly-crystalline graphite are based on the structure of crystallites:

> "*The existing theory of the dimensional changes of polycrystalline graphite relates the dimensional changes of the component crystals parallel and perpendicular to the basal planes denoted by $\Delta X_a/X_a$ and $\Delta X_c/X_c$ respectively to those measured in direction x in the polycrystal.*" (Kelly *et al.*, 1974)

> "*Structural and dimensional changes in polygranular graphite are a function of the crystallites' dimensional change and the graphite's texture.*" (Burchell and Snead, 2007)

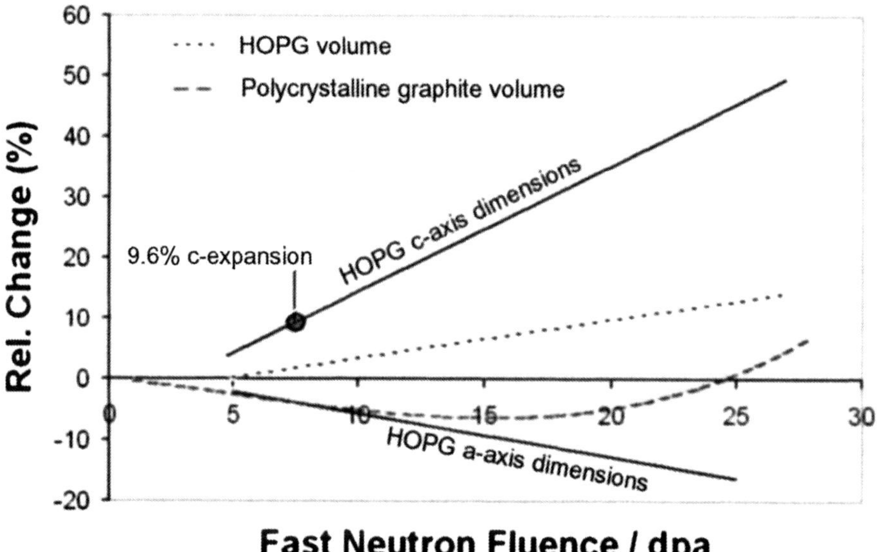

FIGURE 1: Dimensional and volume changes of HOPG irradiated at 500°C (Kelly & Brocklehurst, 1971) compared with volume changes of polycrystalline graphite.

Highly-oriented pyrolytic graphite (HOPG) has been proven to exhibit structure and properties very similar to graphite single crystals. As it can be manufactured with dimensions large enough to be suited as irradiation specimens, HOPG has been irradiated at constant temperatures in the range 270-710°C up to fast neutron fluences of 2.5×10^{22} n/cm^2 (EDND) (Kelly & Brocklehurst, 1971). In the observed fluence range, the dimensions in the a- and c-direction increase linearly with increasing fluence. As an example, Figure 1 shows the dimensional changes at 500°C (Kelly *et al.,* 1974). In addition, the dotted line is representing the volume change calculated from the dimensional changes. Due to c-atoms being displaced from their normal position within the lattice, the volume of HOPG is slowly increasing with increasing irradiation damage with respect to fluence. On the other hand, the volume of polycrystalline graphite exhibits shrinkage instead of expansion in the fluence range below 15 dpa (see Figure 1 – dashed line). This cannot at all be explained by the dimensional changes of the component crystals. Volume shrinkage means densification, and since HOPG cannot be densified another effect compensating the c-axis expansion of crystallites needs to be considered which presumably is related to the porosity of polycrystalline graphite.

MROZOWSKI POROSITY

Porosity is an apparent integral part of polycrystalline graphite. It is introduced by the filler coke; it develops during baking as a consequence of the release of pyrolysis gases, and it evolves during graphitisation and subsequent cooling from graphitisation to room temperature. Depending upon how the pores are formed, their size and geometry can vary in wide ranges.

Porosity developing during the cooling procedure has been described by Mrozowski (1953):

"The structure is formed by heat treatment at high temperature; even if at the highest heat treatment temperature the carbon material should happen to be perfectly dense, as it cools voids will be formed; ... The unavoidable formation of a porosity can be seen directly from, the difference in bulk contraction of a single crystal and of the polycrystalline material, and its magnitude can be easily estimated."

This "Mrozowski" porosity is located adjacent to the crystallites in c-direction, and therefore perfectly suited to accommodate the c-axis growth of the crystallites due to irradiation damage.

Assuming a graphitisation temperature of 3000°C, an average thermal expansion coefficient of 35×10^{-6}/K in the temperature range under consideration, and a final temperature of 500°C (corresponding to a potential irradiation temperature) the size of the Mrozowski shrinkage voids is roughly 8.8% of the crystallite size in c-direction at 3000°C (or 9.6% of the crystallite size at 500°C). These values are upper limits; they may be somewhat smaller if plasticity of the material at the highest temperatures has to be taken into account.

Knowing the amount of expansion of the crystallites that can be accommodated by the Mrozowski voids, Figure 1 gives the neutron fluence at which the *c*-axis expansion of the crystallites should completely fill this kind of accommodation porosity, *i.e.* at an equivalent neutron fluence of about 7.5 displacements per atom (dpa).

DIMENSIONAL AND VOLUME SHRINKAGE

The fact that the a-axis shrinkage is able to outperform the c-axis expansion of the crystallites at fluences below turn-around is ascribed to the accommodation porosity. Once the accommodation porosity has been used up, what will be the implication for further dimensional changes? In the literature (*e.g.*, Burchell *et al.* (2007); Snead *et al.* (2008)), the undisputed answer is that on the one hand "*thermal shrinkage cracks ... initially accommodate the c-direction expansion, so mainly a-direction contraction is observed. Hence, the graphite undergoes net volume shrinkage.*" On the other hand, "*with increasing neutron dose ... the incompatibility of crystal dimensional changes leads to the generation of new porosity, and the volume shrinkage rate falls, eventually reaching zero. The graphite then begins to swell at an increasing rate ... because of the combined effect of c-axis growth and new pore generation. The graphite thus undergoes a volume change 'turnaround' into net growth which continues until the generation of cracks and pores in the graphite, due to differential crystal strain, eventually causes total disintegration of the graphite.*"

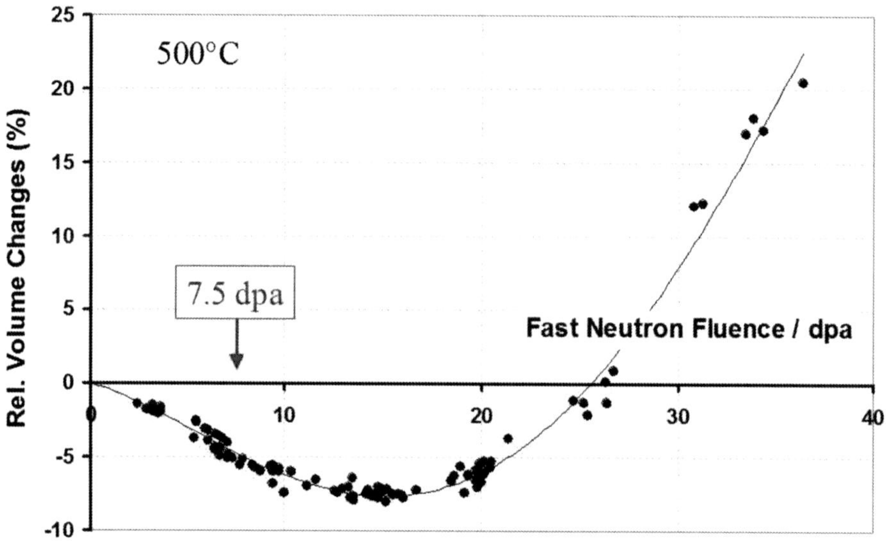

FIGURE 2: Volume changes of an extruded polycrystalline graphite irradiated at 500°C (Haag, 2005). Mrozowski voids are expected to be completely filled at a fluence ≤ 7.5 dpa.

This interpretation ignores that the complete closure of the 'thermal shrinkage cracks' does not coincide with the volume turn-around (see Figure 2). Fundamentally, at the point of complete closure of the Mrozowski porosity, the shape of dimensional changes with increasing neutron fluence should change in the way that the tendency to exhibit expansion would become stronger.

THERMAL EXPANSIVITY CHANGES DUE TO IRRADIATION

After the impurity and the stored-energy problems were solved, until the early 1950s, research activities were directed towards the dimensional stability of graphite under nuclear reactor conditions with the bulk thermal expansion coefficients (CTE) indicating how the graphite should be modified (Kelly, 1978). However, some requirements turned out to be contradictory, *e.g.*, the demand for low coefficients of thermal expansion to enhance dimensional stability which was not compatible with the requirement that nuclear graphite has to be isotropic.

The relationship between dimensional behaviour of polycrystalline graphite under neutron irradiation being associated to the Mrozowski porosity, and thermal expansivity using the same porosity to accommodate crystallite thermal expansion may provide additional insight.

It may be useful to have a closer look to the irradiation behaviour of thermal expansivity in order to find out if an explanation for the open question concerning the dimensional behaviour can be found.

As in the case of dimensional changes, extensive thermal expansivity data are available. No matter where the data have been generated, for isotropic or near-isotropic graphite, they all fit into a pattern similar to Figure 3. After an initial increase, CTE goes through a maximum and then decreases to a level far below the unirradiated value.

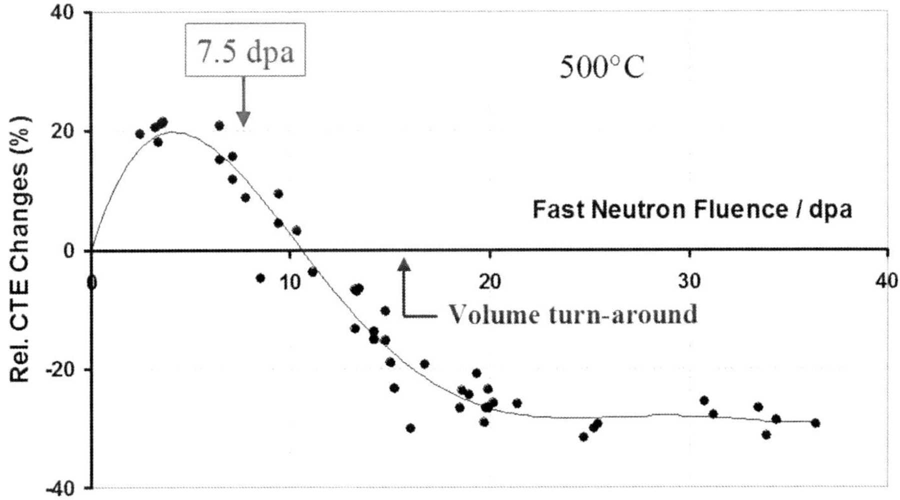

FIGURE 3: Relative changes of the thermal expansion coefficient (20-500°C) of an extruded polycrystalline graphite irradiated at 500°C. Mrozowski voids are expected to be completely filled at a fluence ≤ 7.5 dpa.

The graphite grade and the irradiation conditions in Figure 3 are exactly the same as for Figure 2. It is German data from irradiation tests at 500°C in the High Flux Reactor, Petten, The Netherlands. This time, the broadly accepted explanation of the thermal expansivity behaviour under neutron irradiation is taken from Burchell (1991): "*The coefficient of thermal expansion ... initially increases from its unirradiated value ... and then begins to decrease. The initial increase in CTE can be ascribed to the closure of fine porosity by irradiation induced c-axis crystal expansion. This reduces the available 'accommodation', or internal porosity for the thermally induced c-axis expansion to fill, thus increasing the CTE. Subsequent reductions in CTE are caused by the generation of new porosity with continued neutron damage.*"

If one assumes that the initial increase can be ascribed to the reduction of fine porosity, why does the change of thermal expansivity turn-around at a fluence where the fine porosity has or has almost disappeared (Figure 3)? At first glance, one might think that a decrease in thermal expansivity would call for new accommodation porosity. However, generation of new porosity should not occur before volume turn-around, as it is hard to believe that generation of porosity should go along with shrinkage, or respectively, densification, but again, volume turn-around would require a much higher neutron fluence.

RESULTS FROM OPTICAL MICROSCOPY

Some very old pictures from the early 1970s may give some clues. The first one, Figure 4, is an optical micrograph showing a polished section of a Gilsonite coke graphite as manufactured. Spherical and elliptical coke particles and areas of binder coke can be clearly distinguished. Pores (the black parts of the picture) appear as cracks (from the coking process) within the coke particles, whereas in the binder coke pores are more or less isometric since they are formed by pyrolysis gases which are released during the baking procedure.

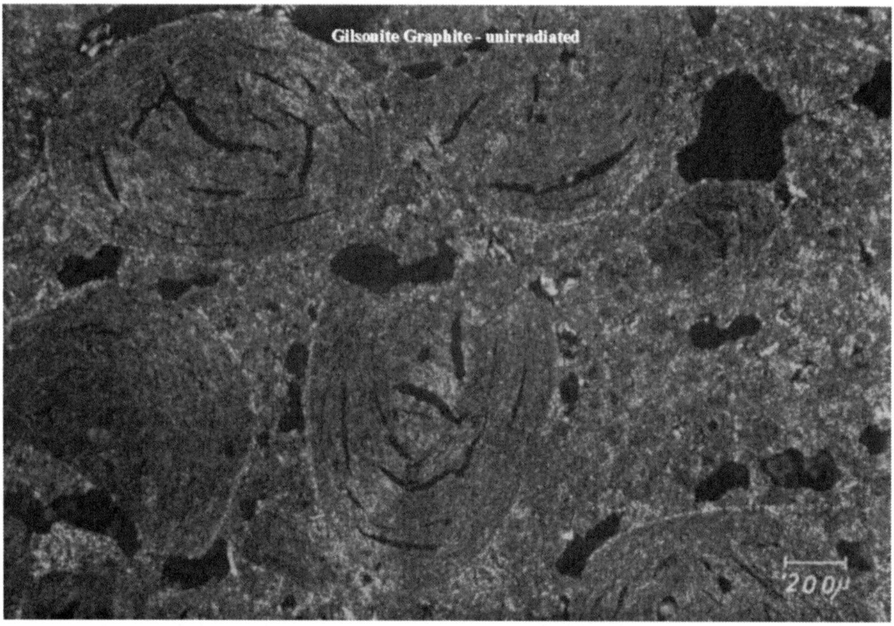

FIGURE 4: Optical micrograph of Gilsocoke graphite – unirradiated.

Figure 5 is showing a polished section of the same Gilsonite coke graphite irradiated to volume turn-around due to a fluence of about 9×10^{21} n/cm^2 (EDND) (the irradiation temperature probably is about 1000°C). The picture has changed very much. While the Gilsonite coke particles are still clearly visible they do not exhibit the typical crack-like porosity anymore, and they seem to have shrunk compared to their size in Figure 4 (same magnification). The porosity in the binder coke has also changed. The former large pores are now smaller, as if they had been too large to disappear completely, whereas pores which were smaller originally have also disappeared. The structure which was irradiated until volume turn-around appears more dense than unirradiated. From Figure 5, it follows that there must be a mechanism that removes voids of any kind not only from the inside of the coke particles, but from the whole bulk material. Even if this mechanism is still unknown, this looks like a major reason for bulk dimensional shrinkage of polycrystalline graphite, which would disprove that the crystallites are the governing species as far as volume shrinkage is concerned.

Figure 6 displays even more progress in structural changes due to irradiation of again the same Gilsonite coke graphite, this time at a fluence of 10.7×10^{21} n/cm^2 (EDND) which is beyond volume turn-around. Even after 14 dpa fast neutron fluence (every C-atom has changed places 14 times!), the typical shape of the coke particles is not lost. Many new pores have been created. Most noticeable are the cracks extending from the periphery to the inside of the coke particles which are obviously caused by tensile stresses. More cracks can be found within the binder coke, *e.g.* connecting existing pores, and thus demonstrating how pores are able to stop crack propagation. The generation of that kind of cracks within and outside the coke particles is well known and has always been related to the turn-around of dimensional changes.

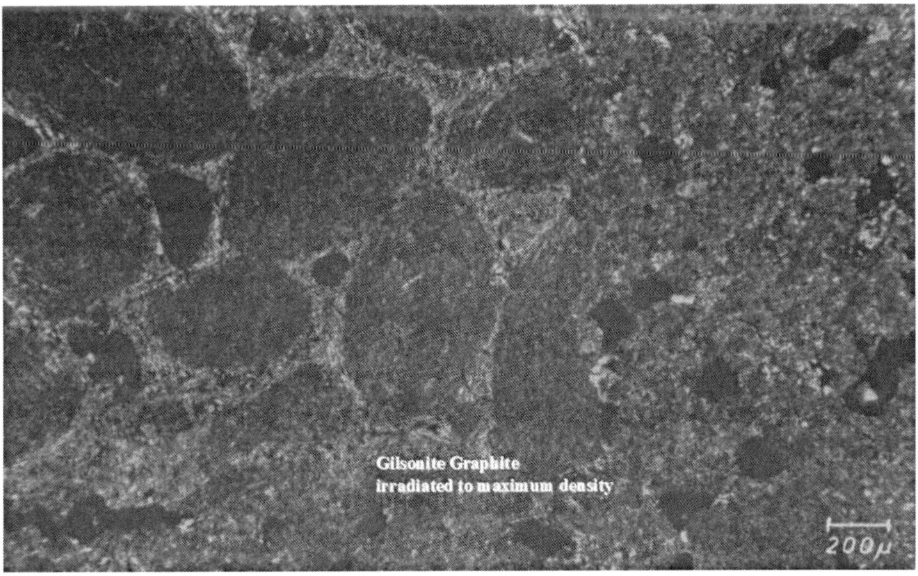

FIGURE 5: Optical micrograph of Gilsocoke graphite – irradiated to volume turn-around.

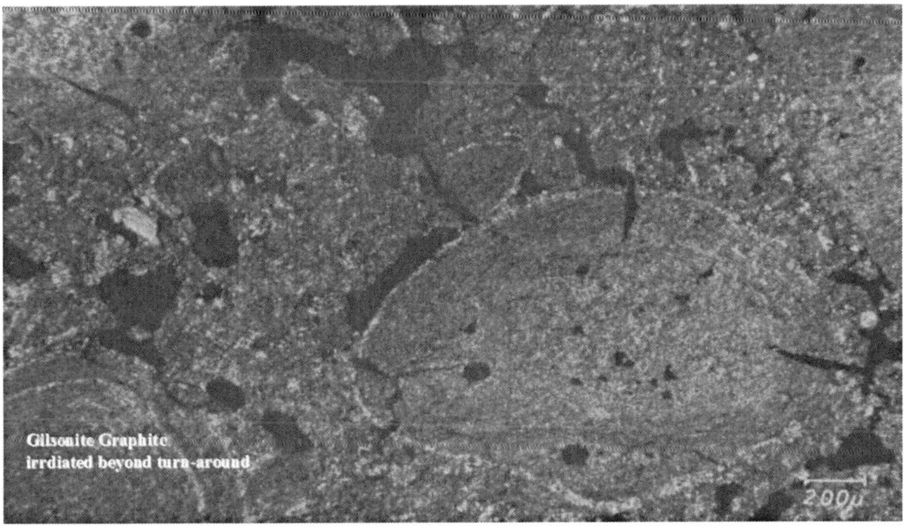

FIGURE 6: Optical micrograph of Gilsocoke graphite – irradiated beyond volume turn-around.

However, there is another kind of pores being created, *e.g.* inside the large Gilsonite particle on the lower right hand side of Figure 6. These voids are spherically shaped and do not obviously result from tensile stress cracking. Their emergence proves that there is still another mechanism creating porosity which also needs explanation. These pores may be the result of coalescing of smaller pores, including those being too small to appear in an optical picture.

If it should be possible that micropores can migrate through the carbon structure to coalesce with others one should consider if they can also accommodate the expansion of crystallites to make thermal expansivity decrease after the Mrozowski pores have been filled up.

CONCLUSION

Models based on appropriate investigations have to be developed in order to explain (1) volume shrinkage in the fluence range from the closure of Mrozowski porosity to volume turn-around; and, (2) the decrease of thermal expansivity after the closure of Mrozowski porosity.

Micrographs show that a huge amount of the porosity is removed from the bulk material due to irradiation causing shrinkage before, later on, new porosity will be created. That new porosity does not only consist of stress cracks, but also of bubble-like pores which look like being formed by coalescing of smaller pores. Micropores, maybe even those from the irradiation-induced shrinkage of crystallites, could provide accommodation volume to explain the decrease in CTE without interfering with the bulk shrinkage due to the removal of porosity. It is obvious that this paper contains some very speculative ideas in order to explain what has not been explained before. However, there are indisputable facts like: (1) the fluence which is necessary to fill the Mrozowski porosity; (2) the unexplained decrease of CTE; (3) the disappearance of a major part of the porosity before turn-around; and, (4) the emergence of bubble-like pores beyond turn-around which have not been discussed before. Therefore advanced investigation techniques should be applied to irradiated graphite in order to find final answers.

References

Burchell, T. D. and Eatherly, W. P. (1991). The effects of radiation damage on the properties of GraphNOL N3M. *J. Nucl. Mat.*, 179-181, 205-208.

Burchell, T. D. and Snead, L. L. (2007). The effect of neutron irradiation damage on the properties of grade NBG-10 graphite. *J. Nucl. Mat.*, 371, 18-27.

Haag, G. (2004). PBMR course and workshop on the use of graphite in PBMR, Centurion, South Africa, 22-25 November, 2004.

Haag, G. (2005). Properties of ATR-2E graphite and property changes due to fast neutron irradiation. Report No. Jül-4183, Forschungszentrum Jülich, October 2005.

Kelly, B. T. (1978). Radiation damage in graphite and its relevance to reactor design. *Progress in Nuclear Energy*, 2, 219-269.

Kelly, B. T. and Brocklehurst, J. E. (1971). High dose fast neutron irradiation of highly oriented pyrolytic graphite. *Carbon*, 9, 783-789.

Kelly, B. T., Brocklehurst, J. E., Ashton, B. W. and Martin, W. H. (1974). *4th Ind. Carbon and Graphite Conf.*, London, 429-437.

Mrozowski, S. (1953). Mechanical strength, thermal expansion and structure of cokes and carbons. *1st Biennial Conference on Carbon*, Buffalo, 31-45.

Snead, L. L., Burchell, T. D. and Katoh, Y. (2007). Swelling of nuclear graphite and high quality carbon fiber composite under very high irradiation temperature. *J. Nucl. Mat.*, 381, 55-61.

Evaluation of Instantaneous Coefficient of Thermal Expansion of PGA Graphite

Martin P.Metcalfe, Nassia Tzelepi and John F. B. Payne
National Nuclear Laboratory, Stonehouse Park, Bristol Road, Stonehouse, Gloucestershire, UK. GL10 3UT.
Email: martin.p.metcalfe@nnl.co.uk

Abstract
A series of coefficient of thermal expansion (CTE) measurements have been undertaken to test the validity of the commonly accepted expressions relating coefficients of thermal expansion of unirradiated graphite over different temperature ranges and the instantaneous coefficient of thermal expansion to the mean coefficient of thermal expansion over a specified temperature range. Results are presented for unirradiated PGA graphite and for irradiated PGA graphite. In the case of irradiated PGA graphite, expressions for evaluating instantaneous values of CTE have been tested at elevated temperatures representative of Magnox reactor operating conditions to support CTE data used to predict thermal stresses in core components.

Keywords
Graphite, coefficient of thermal expansion, temperature dependence

INTRODUCTION

The coefficient of thermal expansion (CTE) is an important parameter in the evaluation of thermal stresses in reactor graphite components. With a dilatometer, the change in sample length is measured over a prescribed temperature range. It is straightforward to use this data to evaluate CTE at any given temperature within that range. For various practical reasons, the measurement temperature range does not always extend to the irradiation temperature. Therefore, a method is needed to extend measured CTE to higher temperatures.

The method employed is based upon CTE data obtained from Materials Test Reactor (MTR) experiments measured over a specified temperature range (20-120 °C) and the use of a correction factor (Martin & Entwisle, 1962) to extend the measured results to the temperatures of interest. The correction factors derived by Martin & Entwisle (1962) were based upon theoretical considerations and measurements made on virgin graphite material. It has been assumed there is no orientation dependence; correction factors apply to anisotropic and isotropic graphite and that they are equally applicable to irradiated material with no limitations on temperature or fluence.

CTE measurements on unirradiated graphite and on graphite removed from high temperature regions of a Magnox reactor core are presented, from which the temperature dependence of instantaneous CTE for Pile Grade A (PGA) graphite has been determined. The test programme objective has been to generate new data that will enable recommendations to be made for CTE application in structural integrity assessments.

BACKGROUND

To a good approximation, the instantaneous CTE, α_i, can be determined from the mean CTE value measured in the range 20-120 °C, $\overline{\alpha}_{20\rightarrow120}$, using the relationship:

$$\alpha_i = A_i \overline{\alpha}_{20\rightarrow 120} + B_i \qquad [1]$$

where A_i and B_i are parameters dependent on temperature. Values for these parameters were proposed by Martin & Entwisle (1962) and later revised by Kelly (1981).

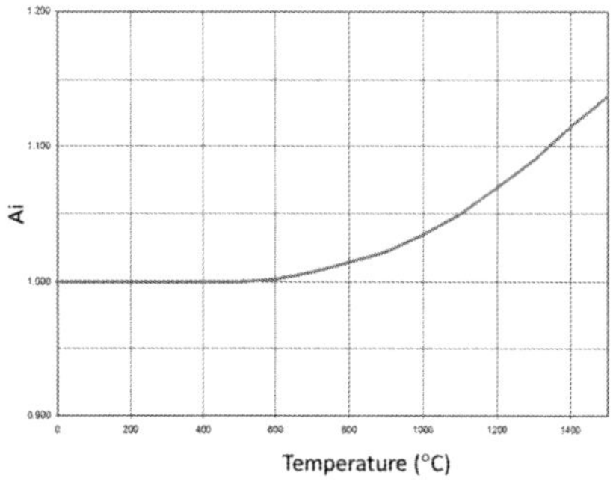

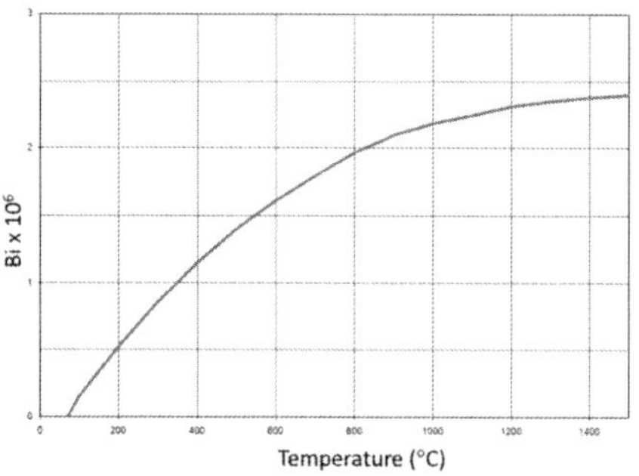

FIGURE 1: Temperature dependence of parameters A_i (left) and B_i (right).

Parameter A_i = 1 for all temperatures of practical interest for Magnox reactors (Figure 1). Values of B_i over the temperature range 0 - 500°C are estimated by linear interpolation (taken to be (-0.05 + 0.0275 T) x 10^{-6}).

The mean CTE over any given temperature range ($T_1 \rightarrow T_2$) can be obtained by integration of the Martin & Entwisle equation and division through by (T_2-T_1). For the case where A_i = 1:

$$\overline{\alpha}_{T_1\rightarrow T_2} = \overline{\alpha}_{20\rightarrow 120} + \overline{B}_{T_1\rightarrow T_2} \qquad [2]$$

MEASUREMENT PROGRAMME

The CTE of unirradiated PGA graphite was measured over the temperature range ambient to 520 °C. For irradiated graphite, specimens trepanned from bricks in the higher temperature positions in the Magnox reactor cores were selected. For irradiated graphite, the CTE was measured over a temperature range determined by its irradiation temperature: ambient to (irradiation temperature + 40 °C). A limit on the upper measurement temperature for irradiated material was set to avoid possible thermal annealing of irradiation damage. This was checked by preceding each high temperature run with a standard temperature run between ambient and 200 °C. If the mean CTE over this temperature range was the same for both the standard and high temperature runs, then thermal annealing was not taking place. Specimens were taken through three temperature cycles with CTE data taken from the average heating and cooling cycles 2 and 3 (the first cycle allowed the specimen to "settle in" within the test apparatus). Unirradiated specimens (12 mm in diameter, 24 mm in length) were prepared such that the CTE could be measured both perpendicular and parallel to the extrusion direction along the cylinder axis. Irradiated specimens were prepared from trepanned cores (12 mm in diameter, 6 mm in length) and measured in the perpendicular direction only across the diameter, with the orientation determined by ultrasonics. Appropriate corrections for silica expansion were made for the two specimen geometries.

METHOD FOR TESTING TEMPERATURE DEPENDENCE OF CTE

For PGA graphite, the fractional length change dL/L_0 from a dilatometer can be accurately described by the quadratic:

$$\frac{dL}{L_0} = a_0 + a_1 T + a_2 T^2 \qquad [3]$$

where T is the temperature in °C and L_0 is the sample length at ambient temperature. The instantaneous CTE is therefore:

$$\alpha = \frac{d}{dT}\left(\frac{dL}{L_0}\right) = a_1 + 2a_2 T \qquad [4]$$

A polynomial fit to the raw dilatometer data permits evaluation of a_1 and a_2 (and a_0). Combining Equations [1] and [4], instantaneous CTE is given by:

$$\alpha_i = A_i \bar{\alpha}_{20\to120} + B_i = a_1 + 2a_2 T \qquad [5]$$

where a_1 and a_2 are fitting parameters for the experimental data and T is the temperature at which the instantaneous CTE is required. Substituting for A_i and B_i using the current methodology (with assumed linear dependence of B_i):

$$\bar{\alpha}_{20\to120} + (-5.00\times10^{-8} + 2.75\times10^{-9}T) = a_1 + 2a_2 T \qquad [6]$$

The temperature dependence $d\alpha/dT$ will be $2a_2$, which is evaluated experimentally and can be compared with the Martin and Entwisle value of 2.75 x 10^{-9} K^{-2}.

EXPERIMENTAL RESULTS

Measurements were made using a Bähr push-rod dilatometer (Type 801) with sample support tubes and push-rods made of fused silica. The temperature program used for the high temperature runs on irradiated graphite together with the associated length changes is illustrated in Figure 2.

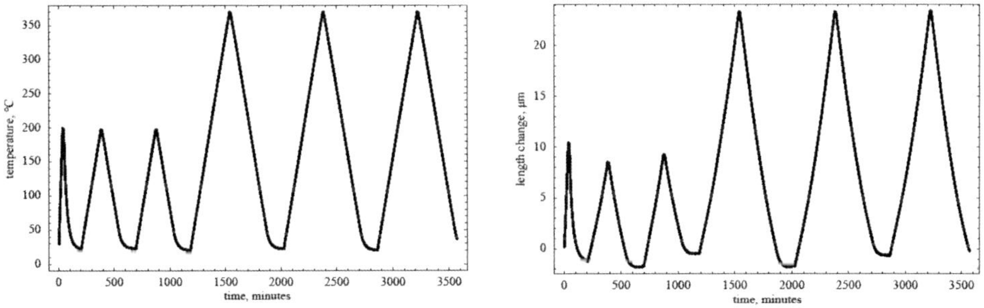

FIGURE 2: Dilatometer traces showing temperature program (left) and associated length changes (right).

It can be seen qualitatively from Figure 2 that the length changes over the three high temperature cycles are broadly similar, confirming there is no thermal annealing. The method described above for evaluating the temperature dependence $2a_2$ was applied to the test data and the results are summarised in Table 1 together with the expected value from the Martin & Entwisle expression.

Graphite condition or source of data	Average CTE temperature dependence ($d\alpha/dT$, 10^{-9} K^{-2})
Martin and Entwisle	2.8
Unirradiated PGA perpendicular	3.2
Unirradiated PGA parallel	3.7
Irradiated brick layer 6	8.1
Irradiated brick layer 7	7.8
Irradiated brick layer 8	7.1
Irradiated brick layer 9	6.8
Irradiated brick layer 10	7.1

TABLE 1: Experimentally measured temperature dependence of CTE.

The results show the temperature dependence of unirradiated PGA graphite to be broadly in agreement with expectation with no strong orientation dependence, given measurement uncertainties. Both CTE and the temperature dependence of CTE are changed by irradiation, by an amount that depends on fluence and irradiation temperature. The fluence dependence of $d\alpha/dT$ is shown in Figure 3 along with a graph showing the trend in CTE for PGA graphite for the same measurement orientation at 250 °C Equivalent DIDO Temperature (taken from historical MTR data).

THEORETICAL CONSIDERATIONS AND DISCUSSION

The prediction of the temperature dependence of the CTE of polycrystalline graphite can be described in terms of single crystal expansion (J. H. W. Simmons, 1965):

$$\alpha_{poly} = \xi\, \alpha_{c\_perp} + (1-\xi)\, \alpha_{c\_para} \qquad [7]$$

where α_{poly} is the CTE of polycrystalline material, ξ is a direction dependent structure factor and α_{c_perp} and α_{c_para} are single crystal CTEs perpendicular and parallel to the crystal hexagonal planes (Figure 4). ξ is orientation dependent, but is commonly assumed to be temperature independent. Since $\alpha_{c\_perp} >> \alpha_{c\_para}$, the CTE of polycrystalline material is dominated by the crystal perpendicular contribution and the directional value of ξ. However, the temperature derivatives of α_{c_perp} and α_{c_para} are of comparable size and have strong but different temperature dependences, so the behaviour of $d\alpha/dT$ is complicated. The overall result, for the temperature range investigated, is that the temperature dependence of α_{poly} is broadly direction independent and so the Martin and Entwisle expression, Equation [7], adequately describes the behaviour of unirradiated graphite.

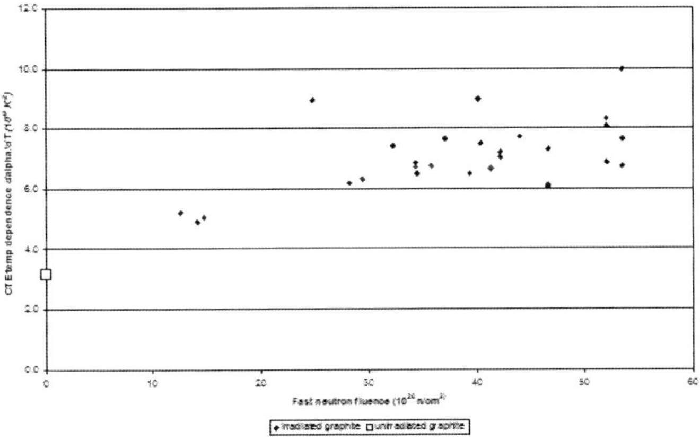

(a)

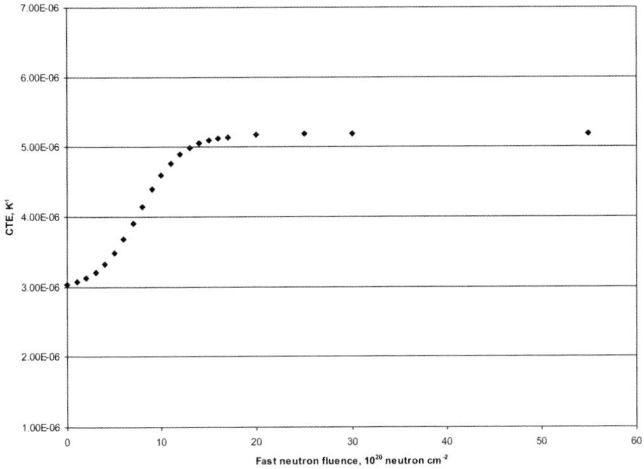

(b)

FIGURE 3: (a) Temperature dependence of CTE; and, (b) CTE as a function of fluence.

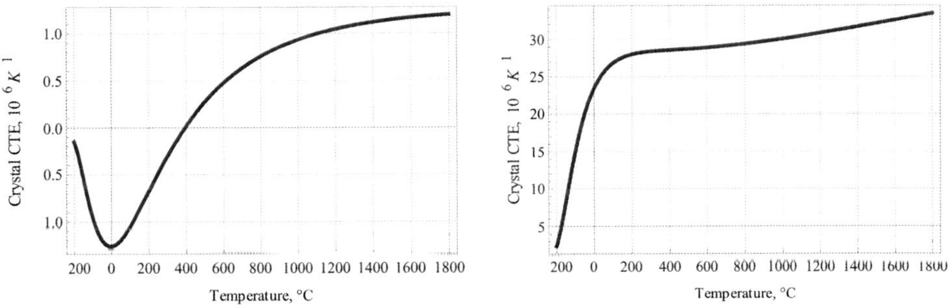

FIGURE 4: Single crystal behaviour parallel (left) and perpendicular (right) to the basal plane (Tsang, Marsden, Fok and Hall (2005)).

For irradiated polycrystalline material, both CTE and the temperature dependence of CTE are influenced by the degree to which microcracks (Mrozowski cracks) in the polycrystal can accommodate expansion. The ξ term is a measure of how much crystal expansion perpendicular to basal planes contributes to the expansion of the agglomerate of differently oriented crystals; the 1- ξ term is a measure of how much crystal expansion parallel to basal planes contributes.

For unirradiated PGA graphite, ξ is small ($\leq$0.15), even for the direction perpendicular to extrusion. Mrozowski cracks accommodate the large CTE perpendicular to the crystal basal planes. In irradiated graphite, crystal expansion perpendicular to crystal basal planes closes Mrozowski cracks so that thermal expansion can no longer be accommodated.

Effectively, ξ increases with irradiation. This is a gradual process as crack widths and rates of closure will differ. At high fluences, where Mrozowski cracks have closed, CTE will be 2-3 times more sensitive to temperature than material with cracks that have not closed. For unirradiated graphite, temperatures of ~2300°C would be needed to completely close these cracks. Much smaller temperatures will close them in irradiated graphite (increasing ξ). This qualitatively explains the observed CTE behaviour of irradiated PGA graphite.

CONCLUSIONS

Martin and Entwisle-type expressions describing the behaviour of polycrystalline graphite broadly agree with measurements on unirradiated PGA graphite over the range ambient-500°C. The CTE of irradiated PGA graphite is not well described by such expressions, so they may not be accurate for other types of graphite when they have been irradiated.

The approach based upon the structure factor qualitatively explains the irradiation-induced increase in the temperature derivative of CTE. Failure to account for the irradiation-induced increase in the temperature derivative of CTE could lead to significant errors when evaluating thermal stresses in graphite components.

Assumed temperature dependence of CTE should be validated against experimental data for irradiated samples of the graphite of interest over the range of fluences and irradiation temperatures of interest. A more detailed microstructural approach is a promising way to reduce the amount of experimental work needed to validate temperature dependence of CTE over the range of fluences and irradiation temperatures of interest.

Acknowledgements

This programme of work, devised and undertaken by the National Nuclear Laboratory, was commissioned and funded by Magnox Ltd. Their support is gratefully acknowledged.

References

Kelly, B T. (1971). Proposals for a revised method for calculating the thermal expansion of graphites and fuel compacts. *UKAEA TRG Memorandum* 5757(S).

Martin, W. H. and Entwisle, F. (1963). Thermal expansion of graphite over different temperature ranges. *J Nucl Mat.*, 10, 1-7.

Simmons, J. H. W. (1965). Radiation Damage in Graphite, *Pergamon Press.*

Tsang, D. K. L., Marsden, B. J., Fok, S. L. and Hall, G. (2005). Graphite thermal expansion relationship for different temperature ranges. *Carbon,* 43, 2902-2906.

A Review of Current Finite Element Models for Irradiation Creep and Failure of Nuclear Graphite

David W. J. Tanner, Adib A. Becker, Wei Sun and Tom H. Hyde
Materials, Mechanics and Structures Research Division, Faculty of Engineering, The University of Nottingham, Nottingham NG7 2RD, UK
Email: david.tanner@nottingham.ac.uk

Abstract
Finite element models currently used for predicting irradiation creep and failure of nuclear graphite are presented. The underlying material properties required for these analyses and the relevant property changes due to irradiation are reviewed, followed by a description of the generation of constitutive equations. Continuum damage mechanics-based approaches for predicting the cracking and failure of graphite components are assessed. The computational implementation of the different models is briefly discussed and some examples from published literature are highlighted. To conclude, areas are identified that shall be addressed in order to develop a unified multi-scale continuum damage material behaviour model for graphite as part of the ongoing EPSRC '*Fundamentals of current and future uses of nuclear graphite*' project.

Keywords
Graphite, irradiation creep, finite element models

INTRODUCTION

As well as acting as a neutron moderator and reflector, graphite is used as a structural component in many gas-cooled fission nuclear reactors. Therefore, the ability to predict the structural integrity of the numerous graphite components that make up a graphite reactor core is important in safety case assessments and reactor core life prediction (Tsang, 2008a).

Nuclear graphite component structural integrity assessments are usually performed using the finite element (FE) method. A complex material model is required in order to model dimensional change, irradiation creep and property changes due to the effects of temperature, irradiation fluence and weight loss. The models used are largely empirical and are based on small samples that constitute material test reactor (MTR) irradiated graphite property data (Tsang, 2006). A comprehensive material property database is required for any graphite component stress analysis.

According to Tsang (2006), a viscoelastic graphite model was first produced in the USA by Chang and Rashid (1970). Later, a model called VIENUS was created by JAERI in Japan for a high temperature test reactor (HTTR) graphite block by Iyoku *et al.* (1991). In the UK, a model for two dimensions and axisymmetric geometries was created using BERSAFE FE code by Harper. This was later developed using ABAQUS FE software by Whittaker (1988) using a User MATerial (UMAT) subroutine to model Gilsocarbon graphite components in two dimensions. More recently, a material model subroutine called MAN UMAT has been developed at The University of Manchester (UK) for nuclear graphite under fast neutron irradiation and radiolytic oxidation conditions. This subroutine can be used together with the ABAQUS FE program to perform three-dimensional, time-integrated, non-linear irradiated graphite stress analyses (Tsang, 2006).

The motivation behind this review was to identify the current state of the art FE models used for nuclear graphite in order to identify gaps and build on existing work to develop a unified multi-scale continuum damage material behaviour model.

MATERIAL PROPERTIES

In order to perform FE stress analyses of nuclear graphite components, the following material properties are required. Two material states exist: the virgin material state, and the irradiated state.

Virgin (Unirradiated) Graphite

The mechanical properties of virgin isotropic graphite are given in Table 1.

Material property	Value
Density	1.81 g/cm^3
Mean coefficient of thermal expansion (CTE)	4.35×10^{-6} (°C)$^{-1}$
Poisson's ratio	0.2
Dynamic Young's modulus (DYM)	10 GPa
Ratio of static Young's modulus (SYM) to DYM	0.84

TABLE 1: Typical virgin material properties of isotropic graphite (Tsang, 2006).

The coefficient of thermal expansion (CTE) is temperature dependent with an increase in temperature producing an increase in the CTE. Poisson's ratio for elastic strain is typically assumed to be 0.2 and to be independent of irradiation, temperature and weight loss.

Stress–strain curves for virgin isotropic graphite subjected to cyclic loading are shown in Figure 1. The macro-deformation is governed initially by elastic shear and then by plastic deformation of individual crystallites, which causes a nonlinear macro response (Tsang, 2006). A large permanent set is evident after the load is removed. Microcrack formation may also be responsible for the nonlinear response as discussed by Hodgkins (2010).

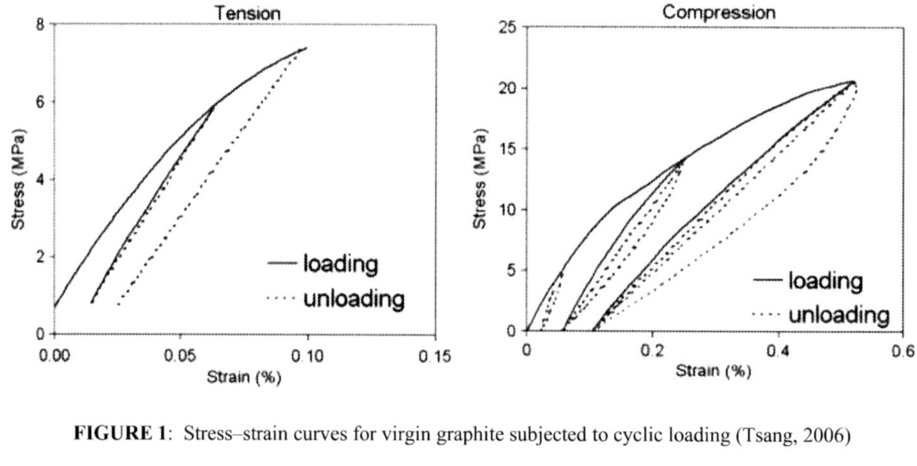

FIGURE 1: Stress–strain curves for virgin graphite subjected to cyclic loading (Tsang, 2006)

An ultrasound method is often used to determine the slope of the stress–strain curve at the origin and this value is referred to as the dynamic Young's modulus (DYM). The static Young's modulus (SYM) is obtained via 'static' tests using strain gauges.

Irradiated Graphite

In a gas-cooled reactor the properties of the graphite components are changed by fast neutron irradiation and, in the case of air or carbon dioxide-cooled reactors (as opposed to helium-cooled), by radiolytic oxidation (Tsang, 2006). The change in dynamic Young's modulus (DYM) of a typical isotropic graphite due to different irradiation fluence levels is shown in Figure 2.

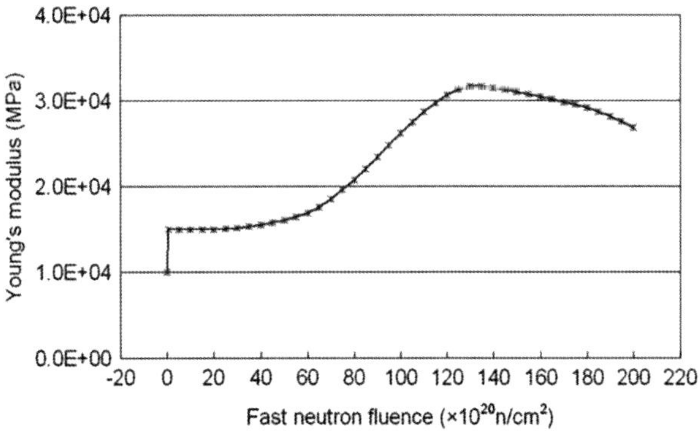

FIGURE 2: Young's modulus variation with fluence for isotropic graphite irradiated at 550 °C (Tsang, 2006).

Fast neutron irradiation fluence also modifies the coefficient of linear thermal expansion (CTE). This change is a function of the irradiation fluence, irradiation temperature and the temperature of the graphite (Tsang, 2006).

Radiation damage of graphite by fast neutron fluence causes crystal growth in the c-axis direction and shrinkage in the a-axis direction. Neutron irradiation of the polygranular graphites used in reactor moderators results in an initial bulk shrinkage at low neutron fluence followed by expansion at higher neutron fluence (Neighbour, 2000). The amount of dimensional change is assumed to be a function of irradiation fluence, temperature and weight loss (due to radiolytic oxidation) (Tsang, 2006). Radiolytic weight loss may amount to greater than 40% of a graphite component throughout the life of a reactor. This weight loss further modifies the Young's modulus.

Graphite in nuclear reactors undergoes irradiation-induced creep, which is covered in the following section. The thermal expansion of graphite is significantly modified by irradiation creep (Kelly, 1991), and so a correction in the CTE is necessary. This is noteworthy since it affects the behaviour of graphite components and their likelihood of cracking during cooling on reactor shut-down.

Sensitivity studies using FE analysis have shown that variations in creep modulus and dimensional change have the most significant effect on stress magnitude (Tsang, 2008a).

IRRADIATION CREEP

Graphite in nuclear reactors undergoes irradiation-induced creep under stress at temperatures where thermal creep is negligible (Burchell, 2010). This is significant since it somewhat relieves otherwise critical internal stresses in graphite moderator blocks, that arise mainly due to the non-uniform dimension change distribution caused by irradiation fluence and temperature gradients (Kelly, 1991).

Irradiation-induced (apparent) creep strain is conventionally defined as '*the difference in dimensions between a stressed and irradiated sample and a sample with the same properties as the stressed sample irradiated unstressed*' (Kelly, 1994). The creep rate is considered to be proportional to stress and practically independent of temperature in the range 300 – 650 °C. Irradiation creep can be characterised by a primary transient stage (which is recoverable) followed by an unrecoverable secondary (steady-state) linear stage (Tsang, 2006). The creep behaviour is assumed to be equal in tension and compression for a given graphite.

The application of constitutive equations for nuclear graphite using standard FE techniques, is further complicated as irradiation creep in graphite is an unusual case since it is considered not to conserve volume (Kelly, 1977), *i.e.* with irradiation, the volume of graphite increases under tensile loading and decreases under compressive loading. Consequently, specially written user coding is required to model the behaviour of irradiated graphite components (Tsang, 2006).

CONSTITUTIVE EQUATIONS AND FE MODEL IMPLEMENTATION

The total strain may be considered as the sum of six different components:

$$\boldsymbol{\varepsilon}^T = \boldsymbol{\varepsilon}^e + \boldsymbol{\varepsilon}^{pc} + \boldsymbol{\varepsilon}^{sc} + \boldsymbol{\varepsilon}^{dc} + \boldsymbol{\varepsilon}^{th} + \boldsymbol{\varepsilon}^{ith} \qquad [1]$$

where $\boldsymbol{\varepsilon}^e$ is the elastic strain, $\boldsymbol{\varepsilon}^{pc}$ is the primary creep strain, $\boldsymbol{\varepsilon}^{sc}$ is the secondary creep strain, $\boldsymbol{\varepsilon}^{dc}$ is the dimensional change strain induced by irradiation, $\boldsymbol{\varepsilon}^{th}$ is the thermal strain, and $\boldsymbol{\varepsilon}^{ith}$ is the interaction thermal strain due to creep. $\boldsymbol{\varepsilon}^{ith}$ is essentially a correction in the CTE due to total creep strain.

As standard, the elastic strain is related to the stress $\boldsymbol{\sigma}$ through Hooke's law of linear elasticity, using the elastic material matrix $\mathbf{D}$, which is a function of Young's modulus and Poisson's ratio:

$$\boldsymbol{\sigma} = \mathbf{D}\boldsymbol{\varepsilon}^e \qquad [2]$$

Although there are various empirically-based models for irradiation creep, the general form that has the most support consists of the sum of the following primary and secondary creep terms (Tsang, 2008b).

The primary creep strain is defined as:

$$\boldsymbol{\varepsilon}^{pc} = a\varphi e^{-a\gamma} \int_0^{\gamma} \mathbf{D}_c \boldsymbol{\sigma} e^{a\gamma'} d\gamma' \qquad [3]$$

where φ is a temperature dependent parameter, a is a constant, γ is the irradiation fluence and $\mathbf{D}_c$ is the creep material matrix.

And the secondary creep strain is defined as:

$$\boldsymbol{\varepsilon}^{sc} = \xi \int_0^{\gamma} \mathbf{D}_c \boldsymbol{\sigma} d\gamma' \qquad [4]$$

where ξ is a temperature dependent parameter. $\mathbf{D}_c$ is a function of the creep modulus and creep Poisson's ratio (often assumed to be equal to the elastic Poisson's ratio). The initial (unirradiated) creep modulus is usually assumed to be the same as the unirradiated elastic modulus and is similarly modified by irradiation (Davies, 2008; Tsang, 2008a). Alternative equations for irradiation-induced creep of graphite may be found in Burchell (2010).

For a specified temperature, the dimensional change rate subject to irradiation γ and weight loss ϕ is given by:

$$\frac{d\boldsymbol{\varepsilon}^{dc}}{d\gamma} = \mathbf{h}(\gamma, \phi) \qquad [5]$$

The thermal strain, as standard, is defined as:

$$\boldsymbol{\varepsilon}^{th} = \bar{\boldsymbol{\alpha}}_{(T_{ref}\text{-}T)}\left(T\text{-}T_{ref}\right) \qquad [6]$$

where $\bar{\boldsymbol{\alpha}}_{(T_{ref}\text{-}T)}$ is the mean CTE between a reference temperature T_{ref}, and temperature T.

The incremental forms of the strain components, and the Jacobian matrix $\partial\Delta\boldsymbol{\sigma}/\partial\Delta\boldsymbol{\varepsilon}^T$, are required by the FE code. A method for determining the interaction thermal strain, which cannot be calculated explicitly due to its dependence on the total creep stain, may be found in Tsang (2006).

To perform an FE stress analysis for nuclear graphite under irradiation, four different sets of input data are required: an FE mesh of the component geometry, field variable data such as irradiation fluence, temperature and fractional weight loss spatial distributions and variations with time, the virgin material properties and the irradiated material properties. The material property data required are the coefficient of thermal expansion, the dimensional change, Young's and creep moduli, and elastic and creep Poisson's ratios (Tsang, 2008a). Irradiation fluence within a graphite moderator brick can be predicted using standard reactor physics codes by Monte Carlo simulation. The temperature distribution can be calculated using FE heat transfer analysis, incorporating nuclear heat generation, conduction throughout bricks and between bricks via interlocking keys, and convective heat transfer. The weight loss distribution can be predicted by extrapolating reactor sampling data.

A whole reactor core is too large a structure and contains too many interactions for a conventional FE analysis. Instead, super-element models and user element subroutines or substructure modelling techniques along with parallel programming may be used to keep the computer memory requirements and computational times to a manageable size.

FAILURE MODELS

Graphite is a brittle material and its failure strength depends on component size, shape and mode of loading. For specimens of a similar size, graphite is stronger in compression then bending, and stronger in bending than in tension. In regions of high stress concentration, components have a much higher failure stress than predicted through small-scale testing. Thus it is difficult to predict the failure of a graphite component from small specimen tests even without considering the added complexity of irradiation damage. Currently, a large amount of full-scale testing is necessary to determine failure stresses for all the different loading modes. The tests are performed on unirradiated components and then modifications for the effects of irradiation and oxidation are made. Therefore, there is a need for a more rigorous and flexible method of predicting failure in graphite components, without requiring full-scale tests (Marsden, 2005).

Failure models based on Griffith's theory and strain energy criteria have been used for graphite and Burchell (1996) combined a fracture mechanics approach with a microstructurally based description of fracture. The tensile, compressive and flexural strength of graphite has been shown to increase with neutron fluence before peaking and declining, following the trend in Young's modulus (Price, 1979). This is consistent with the Griffith-Irwin model for the fracture of brittle materials.

Continuum Damage Mechanics

Failure models based on continuum damage mechanics (CDM) have been developed for nuclear graphite and implemented using FE analysis. These models implicitly include irreversible microstructural changes such as micro-cracking via damage state variables. The brittle damage constitutive equation developed by Chow and Yang (1991) was used to simulate the non-linear macro deformation and failure of graphite in Kaji (2001). A damage surface similar to the yield function in the conventional theory of plasticity was introduced.

Zou (2004) coupled the effects of mode-I and mode-II loading in an FE CDM model for nuclear graphite. An interface element with zero thickness was introduced into the continuum solid at positions where potential crack surfaces may form. Since nuclear graphite appears to fail under the maximum principal stress (mode-I), it was proposed that a linear stress analysis should be performed to determine the location and direction of the maximum principal stress so that the likely cracking site and crack growth path could be introduced into the FE mesh. Within the interfacial constitutive law, which relates tractions and displacements across the interface, a damage parameter was introduced to take account of the effects of damage such as micro-cracking. A damage surface was constructed using both the conventional stress-based and fracture mechanics-based failure criteria. Therefore, this CDM model can predict both the initiation and propagation of cracks in components with high stress concentration or even a stress singularity (Marsden, 2005). The initiation and development of damage was controlled by a stress-based failure function, and crack surface formation was determined by a fracture mechanics-based failure function. Failure predictions for various virgin graphite components, under three-point bending, four-point bending and brick slices under internal and external stresses provided very good agreement with experimental data (Zou, 2006). To use the CDM model to simulate the failure of nuclear graphite under irradiation and radiolytic oxidation, the values in the stress-based and fracture mechanics-based criteria will need to change as for the other material properties mentioned previously.

FUTURE WORK

Accurate lifetime assessment of nuclear graphite components requires an understanding of the material behaviour over a range of length scales from the microscopic level of irradiation induced defect effects, through the mesoscopic level of dislocation dynamics, up to cracking and failure on the macroscopic level. The challenge is to determine the long-term damage evolution of graphite components under various conditions. To achieve this, it is necessary to study phenomena such as creep, shrinkage and swelling, fracture, oxidation, any impurity effects and the correlation between microstructural features and bulk behaviour, as well as property changes due to irradiation (Samaras, 2007). This task involves kinetic Monte Carlo simulations at one end of the length scale informing FE continuum models at the other.

Radiation creep is arguably the most important of the radiation effects that occur in graphite. Recently, the breakdown of the currently accepted models for creep at high temperatures and high neutron doses has been reported (Burchell, 2010). In general the development of detailed theoretical models of radiation creep in graphite derived from atomic and microstructural scale mechanisms has not been carried very far. The models have been mostly based on empirical or semi-theoretical approaches and there is potential to use high resolution electron microscopy along with atomistic simulations to resolve the microscopic mechanisms of irradiation creep in graphite and help develop mechanistic models which are suitable for input into FE stress analyses (Burchell, 2010). Models for the multiaxial irradiation creep behaviour of graphite –essential for accurate component failure predictions, require attention since data has been generated under uniaxial loading and transferability to the multiaxial stress state is not clear (Knott, 2010). For example, the creep strain "Poisson's" ratio needs to be confirmed for multiaxial cases (Wang, 2008).

Graphite used for moderator bricks in UK gas-cooled reactors consists of relatively large graphitised filler particles dispersed in a matrix of finer graphite crystallites defined as 'flour' (crushed filler) and graphitised pitch binder. Pores, that arise due to gas evolution and inter-crystalline accommodation cracking due to thermal strains resulting from the baking and graphitisation processes during manufacture, exist in both the filler particles and the matrix (Hodgkins, 2010). Therefore, an important area of ongoing research concentrates on how this microstructure (and its evolution) governs the macro (bulk) material behaviour, which can be studied via multiphase FE models (*e.g.* Berre, 2008a). Similarly, quasi-brittle fracture models such as the smeared-crack failure model, originally developed for conveniently describing the deterioration process caused by micro-cracking within concrete materials, are now being used to model failure of graphite (Berre, 2008b). An improved mechanistic understanding of the effects of irradiation on graphite microstructural property relationships will also enable a reduction in the amount of costly MTR programmes and reduce the reliance on the empirical fitting to test data (Marsden, 2010).

A unified multi-scale continuum damage material behaviour model for graphite, in conjunction with a discrete crack approach to represent final fracture, will be developed in the EPSRC '*Fundamentals of current and future uses of nuclear graphite*' project.

REFERENCES

Berre, C., Fok, S. L., Marsden, B. J., Mummery, P. M., Marrow, T. J. and Neighbour, G. B. (2008a). Microstructural modelling of nuclear graphite using multi-phase models. *Journal of Nuclear Materials*, 380(1-3), 46-58.

Berre, C., Fok, S. L., Mummery, P. M., Ali, J., Marsden, B. J., Marrow, T. J. and Neighbour, G. B. (2008b). Failure analysis of the effects of porosity in thermally oxidised nuclear graphite using finite element modelling. *Journal of Nuclear Materials*, 381(1-2), 1-8.

Burchell, T. D. (1996). A microstructurally based fracture model for polygranular graphites. *Carbon*, 34(3), 297-316.

Burchell, T. D., Murty, K. L. and Eapen, J. (2010). Irradiation induced creep of graphite. *Journal of the Minerals, Metals and Materials Society*, 62(9), 93-99.

Chang, T. Y. and Rashid, Y. R. (1970). Viscoelastic response of graphitic materials in irradiation environments. *Nuclear Engineering and Design*, 14(2), 181-190.

Chow, C. L. and Yang, F. (1991). On one-parameter description of damage state for brittle material. *Engineering Fracture Mechanics*, 40(2), 335-343.

Davies, M. A. and Bradford, M. (2008). A revised description of graphite irradiation induced creep. *Journal of Nuclear Materials*, 381(1-2), 39-45.

Hodgkins, A., Marrow, T. J., Wootton, M. R., Moskovic, R. and Flewitt, P. E. J. (2010). Fracture behaviour of radiolytically oxidised reactor core graphites: a view. *Materials Science and Technology*, 26(8), 899-907.

Iyoku, T., Ishihara, M. and Shirai, H. (1991). Development of thermal/irradiation stress analytical code 'VIENUS' for HTTR graphite block. *Journal of Nuclear Science and Technology*, 28(10), 921-931.

Kaji, Y., Gu, W., Ishihara, M., Arai, T. and Nakamura, H. (2001). Development of structural analysis program for non-linear elasticity by continuum damage mechanics. *Nuclear Engineering and Design*, 206(1), 1-12.

Kelly, B. T. (1991). A review of irradiation creep in reactor graphite. *The Status of Graphite Development for Gas Cooled Reactors, Proceedings of IAEA Specialists' Meeting*, Ibaraki, Japan, 200-204.

Kelly, B. T. and Brocklehurst, J. E. (1977). UKAEA Reactor Group studies of irradiation-induced creep in graphite. *Journal of Nuclear Materials*, 65(0), 79-85.

Kelly, B. T. and Burchell, T. D. (1994). The analysis of irradiation creep experiments on nuclear reactor graphite. *Carbon*, 32(1), 119-125.

Knott, J. F. (2010). Engineering structural integrity assessment: Current goals. *Tenth International Conference on Engineering Structural Integrity Assessment (ESIA10), Manchester, UK: EMAS*, 3-14.

Marsden, B. J., Fok, S. L. and Li, H. (2005). Irradiation behaviour and structural analysis of HTR/VHTR graphite core components. *18th International Conference on Structural Mechanics in Reactor Technology (SMiRT 18)*, Beijing, China, 38-47.

Marsden, B. J., Hall, G. N., Jones, A. N. and Marrow, T. J. (2010). The life cycle of nuclear graphite in Generation IV high temperature reactors. *Tenth International Conference on Engineering Structural Integrity Assessment (ESIA10), Manchester, UK: EMAS*, 193-212.

Neighbour, G. B. (2000). Modelling of dimensional changes in irradiated nuclear graphites. *Journal of Physics D: Applied Physics*, 33(22), 2966-2972.

Price, R. J. (1979). Strength of irradiated graphite: A review. *IAEA Specialists Meeting on Mechanical Behaviour of Graphite for HTRs*, Gif-sur-Yvette, France, 69-75.

Samaras, M., Hoffelner, W. and Victoria, M. (2007). Modelling of advanced structural materials for GEN IV reactors. *Journal of Nuclear Materials*, 371(1-3), 28-36.

Tsang, D. K. L. and Marsden, B. J. (2006). The development of a stress analysis code for nuclear graphite components in gas-cooled reactors. *Journal of Nuclear Materials*, 350(3), 208-220.

Tsang, D. K. L. and Marsden, B. J. (2008a). Constitutive material model for the prediction of stresses in irradiated anisotropic graphite components. *Journal of Nuclear Materials*, 381(1-2), 129-136.

Tsang, D. K. L., Marsden, B. J., Vreeling, J. A. and van der Laan, J. (2008b). Analyses of a restrained growth graphite irradiation creep experiment. *Nuclear Engineering and Design*, 238(11), 3026-3030.

Wang, H. and Yu, S. (2008). Uncertainties of creep model in stress analysis and life prediction of graphite component. *Nuclear Engineering and Design*, 238(9), 2256-2260.

Whittaker, H. J. (1988). Influence of creep Poisson's ratio on irradiation induced self stress in graphite. *4th UK ABAQUS User Group Conference.*

Zou, Z., Fok, S. L., Marsden, B. J. and Oyadiji, S. O. (2006). Numerical simulation of strength test on graphite moderator bricks using a continuum damage mechanics model. *Engineering Fracture Mechanics*, 73(3), 318-330.

Zou, Z., Fok, S. L., Oyadiji, S. O. and Marsden, B. J. (2004). Failure predictions for nuclear graphite using a continuum damage mechanics model. *Journal of Nuclear Materials*, 324(2-3), 116-124.

Quasi-Brittle Fracture Concepts to Improve Structural Integrity Assessments of High Oxidation Weight Loss Graphite Components

Andrew D. Hodgkins, Mahmoud Mostafavi, Peter E. J. Flewitt, Malcolm R. Wootton and Robert Moskovic
Serco TCS, Walton House, Faraday Street, Birchwood, Warrington, Cheshire, WA3 6GA.
Email: Andrew.hodgkins@serco.com

Abstract
Quasi-brittle fracture behaviour is characterised by the shape of the load-displacement curve and the associated mechanisms. The curves commonly exhibit non-linearity in the rising part followed by a progressive loss of load carrying capacity post peak. The behaviour of quasi-brittle materials is subject to wide variations, in the load-displacement response, both tensile and compressive, which are associated with stochastic and deterministic size effects, component geometry, load regime and load rate. The behaviour is observed in materials with brittle, heterogeneous, porous microstructures such as concretes, porous ceramics, reactor core graphites and many geological materials. Non-linearity up to the peak load is associated with the formation of distributed micro-cracks that accommodate strain and modify the ability of the material to store strain energy. In this paper, we explore how experimental observations from a range of quasi-brittle materials may help to provide improved understanding of how the microstructure can influence the fracture behaviour of reactor core moderator graphite by modifying the elastic strain energy storage capacity. This is briefly discussed with respect to the impact on the structural integrity of irradiated reactor core graphites.

Keywords
Quasi-brittle, fracture, microstructure

INTRODUCTION

Quasi-brittle fracture behaviour is a complex interaction of different effects that produce different modes of failure under different load conditions. In addition, the load-displacement characteristics are commonly subject to wide variations which may, in some cases, appear to contain inconsistencies. Both effects are associated with differences in microstructure, which can include inherent variability of a given material (Hodgkins *et al.*, 2010). The quasi-brittle class is formed from materials as diverse as concrete and trabecular bone, ceramic foams and graphite. As a result, it is important to improve understanding of quasi-brittle fracture behaviour.

The objective of this paper is to explore the capacity of the microstructure to influence cracking behaviour in a range of materials, which have quasi-brittle characteristics. However, the influence of the microstructure has to be considered in combination with other variables that can contribute to the behaviour and help to explain fracture behaviour over different length scales.

The first part of the paper summarises current understanding of quasi-brittle fracture behaviour. Part two is an examination of materials with typical quasi-brittle microstructures with emphasis on the features that influence damage accumulation and crack propagation. The third part of the paper examines the capacity of these various microstructures to accommodate damage and to influence strain energy storage.

CHARACTERISTICS OF QUASI-BRITTLE LOAD-DISPLACEMENT BEHAVIOUR

The load-displacement curve for a linear elastic material in tension consists of a continuous rise to the peak load followed by a sudden sharp drop at fracture (Figure 1a). The load-displacement curve for an elastic-plastic material contains significant non-linearity in the rising part of the curve caused by the effects of plasticity prior to fracture (Figure 1c). Typically, the load-displacement curve for a quasi-brittle material loaded in tension, bending and compression is characterised by two key features; a non-linear rise up to the peak load, followed by progressive loss of load carrying capacity after the peak (Figure 1c).

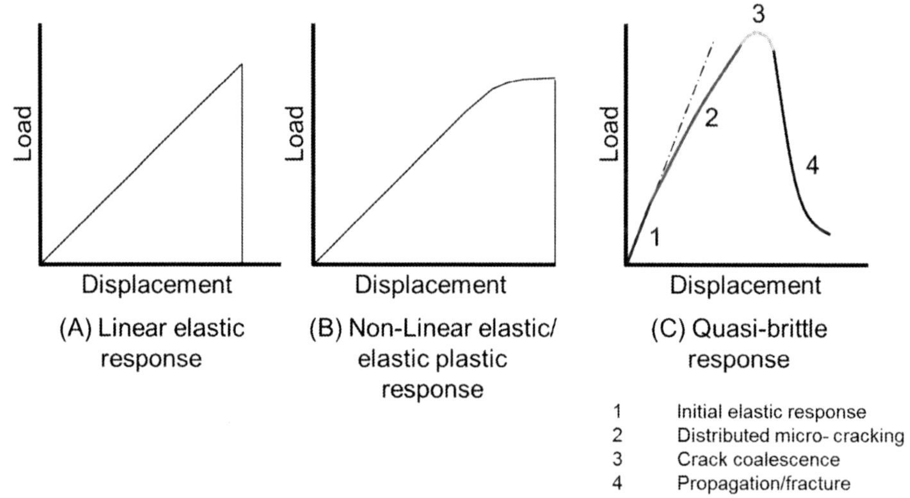

FIGURE 1: Schematic load-displacement curves for each of the three major classes of fracture behaviour. The key beneath the quasi-brittle curve relates the behaviour to mechanisms understood to be responsible.

Non-linearity in the rising part of the curve is considered to be associated with the formation of distributed micro-cracks (Hodgkins, 2006). Close to the peak load, the rate of damage accumulation may be observed to increase. This is associated with stress/strain localisation, which results in the formation of a locally dominant macro-crack. The drop in the load that characterises the post peak behaviour is linked to macro-crack propagation.

Pre-peak non-linearity and post peak softening are associated with mechanisms that are fundamental to the microstructure. The classic quasi-brittle load-displacement curve shown in Figure 1c is associated with small test specimens; the fracture behaviour of very large structures may not exhibit the non-linearity associated with quasi-brittle behaviour.

The relation between fracture behaviour and size in quasi-brittle materials is shown graphically in Figure 2 (Hodgkins *et al.*, 2010). Here the fracture strength is plotted against the specimen/component size, D to show the relation between the size of the stressed volume, the nominal strength and the changes in the statistical distribution. Broadly the effects may be divided into two parts; a stochastic component to explain the scatter and a deterministic component to explain the progressive reduction in the nominal strength. Two notable changes are highlighted in the statistical distribution, of the strength:

1. A gradual reduction in the standard deviation at small length scales, which is associated with the influence of the local variations in the resistance of the microstructure.
2. A change in the form of the distribution (Normal to Weibull), which indicates a fundamental shift in the failure mode; the probability of fast fracture at initiation increases related to the change to the width of the distribution.

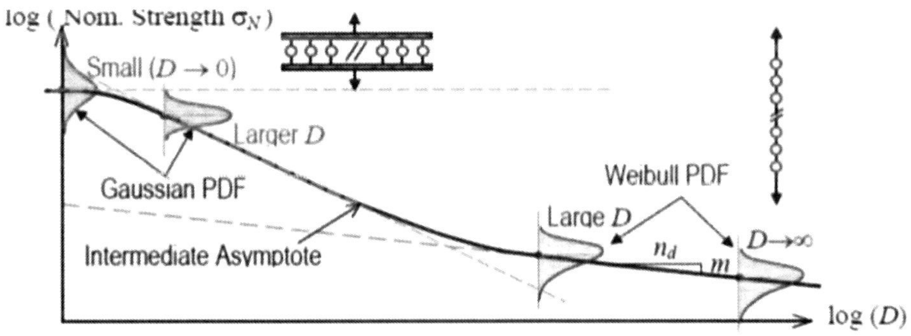

FIGURE 2: Schematic curve showing the influence of the size of the stressed volume on the predicted fracture behaviour.

The reduction of strength associated with larger test specimens or structures is associated with the increased capacity to store strain energy. In addition, to the length-scale dependency, mentioned briefly above, the behaviour is influenced by geometry and stress field effects (Becker, 2011). This dependency may also be linked to microstructure variability. Every test specimen is comprised of a unique volume of microstructure that reacts differently to the applied load conditions. The differences result in variations in damage accumulation, which impact on strain energy. Evidence for the influence of local microstructure variability can be seen by comparing the load-displacement curves for nominally identical specimens. Figure 3 shows the load-displacement curves, for two specimens of PGA graphite, loaded in 4-point bend. Based on the shape of the two curves the values obtained for the elastic modulus and the fracture strength exhibit significant differences. The work of fracture, defined by the area beneath the curve up to the peak load is equally dissimilar for the two specimens.

A change to the way the specimens are loaded can produce an even greater effect. Tensile or compressive loads produce a stressed volume that is uniformly loaded. As a consequence, the microstructure is sampled uniformly, whereas bending loads produce a stress/strain gradient through the stressed volume. As a result the microstructure is not uniformly sampled. As a consequence damage accumulation and spatial sampling within the stressed volume is significantly changed (Anderson, 1994).

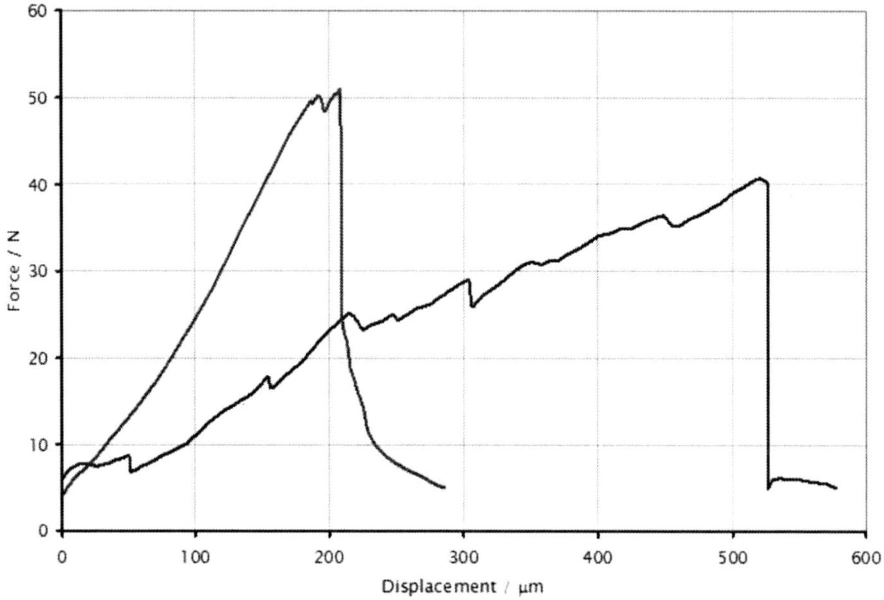

FIGURE 3: Load-displacement curves showing the variation in the behaviour of nominally identical specimens of PGA graphite in 4 point bending.

It is particularly difficult to measure the fracture properties of quasi-brittle materials. Fracture toughness tests developed for elastic or elastic-plastic materials produce values for the fracture toughness that are strongly bound to test specimen geometry. In addition, values for the work of fracture, G and J rely heavily on assumptions about the way energy is absorbed and the shape of the crack front. There have been numerous studies that show the formation of distributed micro-cracks in a constrained process zone around the crack tip. These absorb energy and alter the elastic properties around the crack tip. More recently, studies using X-ray tomography have been used to demonstrate the assumption of a straight crack front may be invalid in all but the smallest specimens (Hodgkins, 2006). The sequence of tomographic images in Figure 4 show the progressive growth of a macro-crack following a successive series of load cycles in a load-unload test on a compact tension specimen of un-irradiated Gilsocarbon graphite (Hodgkins, 2006). The images appear to show a very irregular crack front, however, due to the interaction between the growing crack and the porosity it is not certain that the surface observed in the image was all produced by the loading in the test.

QUASI-BRITTLE MICROSTRUCTURES

The quasi-brittle class of materials is classically described by reference to their brittle heterogeneous microstructures, which often include a significant volume fraction of porosity. Materials within the class are commonly formed by fusing hard ceramic aggregates/grains at elevated temperature and pressure (Hodgkins, 2006). The bond between grains is influenced by grain surface texture, contact area, temperature, pressure and time. The grain size distribution is also recognised to be important because of its influence on aggregates/grains packing (Tucker, 2001). Densification and bonding may be improved by addition of a liquid phase during formation. Under the influence of heat and pressure the liquid phase transforms into a matrix that wraps around the aggregates/grains to improve the contact.

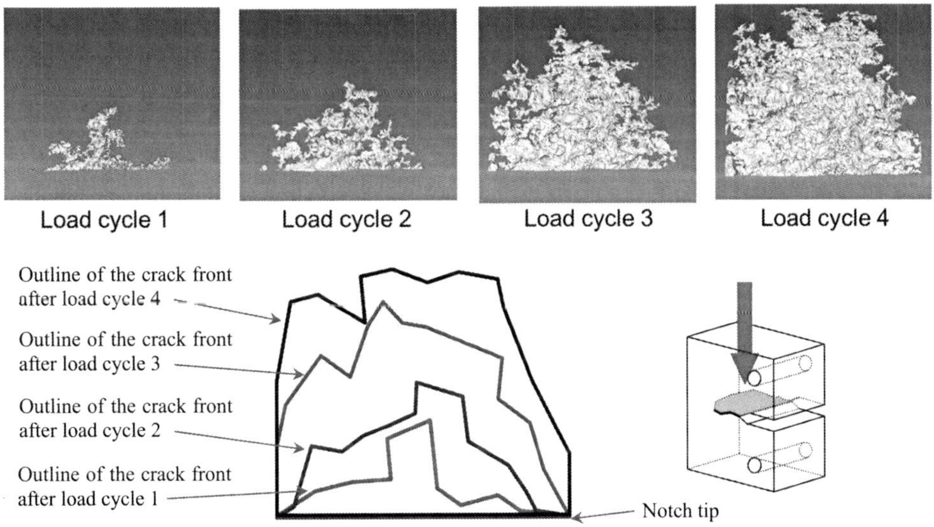

FIGURE 4: Tomography images showing the apparent growth along the crack front produced by four successive load cycles in a compact tension specimen of un-irradiated Gilsocarbon graphite.

Where the aggregates are dispersed within a continuous framework of matrix material the load transfer characteristics and the stiffness may be different to regions where the aggregates are densely packed to form a rigid framework. Geologists refer to this packing as the fabric of the microstructure; densely packed grains forming a continuous framework are commonly termed a grain supported fabric (Tucker, 2001). Where the aggregates are dispersed in a continuous framework of the matrix material the fabric is referred to as "matrix supported" (Tucker, 2001).

Graphite is produced by heating a mixture of dried, calcined coke particles and liquid pitch to temperatures around 2600°C – 3000°C (Hodgkins, 2006). The amount of porosity in the final product is controlled by the addition of liquid pitch, which is impregnated into the structure, at elevated temperature following baking. As a result, the microstructure of graphite shares many of the characteristics and features common to the quasi-brittle class. Relatively large aggregates/filler particles dispersed within a matrix, which forms from liquid pitch (Hodgkins, 2006).

The porosity observed in graphite is commonly divided into two categories. Networks of gas evolution porosity define the pathways used by gases to escape from the green structure (Hodgkins *et al.*, 2010). They form and grow principally in the matrix phase because the aggregates are often relatively impervious to gas flow. Gas evolution porosity is not uniformly distributed through the matrix; it is more commonly observed in 'patchy regions' that mark out the pathways to the surface (Hodgkins *et al.*, 2010). As shown in Figure 5, this patchy distribution is not unique to graphite; many other quasi-brittle materials, such as concrete, exhibit similar features (Diamond and Huang, 1998). The thermal cycle introduces stresses due to the differential contraction between differently oriented crystallographic planes; the stresses are alleviated by the formation of micro- cracks. The dimensions of the cracks approximate to the scale of the crystallites.

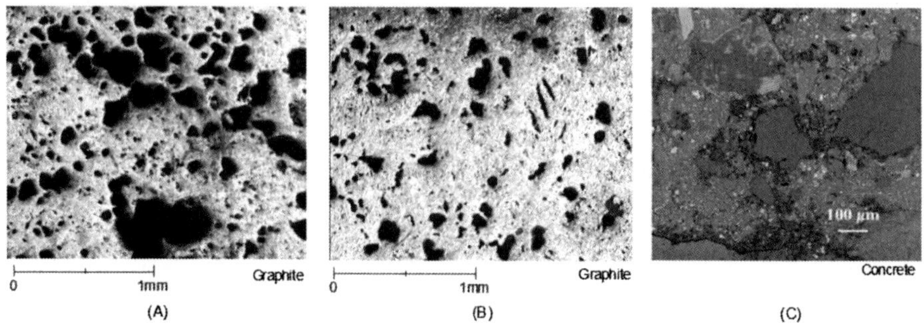

FIGURE 5: Microscopy images showing regions where the density of the porosity is patchy. This is shown in un-irradiated Gilsocarbon graphite in (A) & (B) and in concrete in (C).

THE INFLUENCE OF THE MICROSTRUCTURE ON DAMAGE ACCUMULATION AND STRAIN ENERGY STORAGE

A linear elastic continuum stores elastic strain energy in proportion to; the applied load, the elastic modulus, the size of the stressed volume and the stress/strain gradient through the stressed volume (Ewalds, and Wanhill, 1989). However, in quasi-brittle structures the stressed volume is not a continuum; loads are transferred through pathways of variable stiffness that form within the fabric of the microstructure. The stiffest pathways attract the greatest load and store the most energy. Under displacement control, when the weakest link fractures the pathway ceases to transfer load; the strains are transferred to alternative, less stiff, pathways. As a consequence, the accumulation of micro-crack damage progressively reduces the global stiffness allowing the strains to be accommodated and limiting strain energy storage. Continued loading causes the micro-cracks to coalesce to form locally dominant macro-cracks. At longer crack lengths a process zone of distributed micro-cracks is observed to form around the crack tips (Bazant, 1998). Consequently the shape of the load-displacement curve and the mode of fracture are governed by the distribution of the resistance of the microstructure in the stressed volume.

Studies in a range of quasi-brittle materials show that the individual phases behave differently under load. In concrete (aggregate, matrix and porosity), for example, the interfaces are commonly weak relative to the aggregate particles and the crystallites that comprise the matrix (Boel *et al.*, 2007; Diamond & Huang, 2001). However, in all but the very smallest structures (those in which the stressed volume is relatively small compared to the size of the aggregates) there is likely to be a very large number of interfaces thereby resulting in a significant number of potential micro-cracks initiation sites. Once a crack has initiated it may propagate rapidly through the interface region without the need to significantly increase the load. Cracks may deflect, bifurcate or arrest due to the interaction with the microstructure interface. It is these mechanisms that lead to the characteristically rough fracture surfaces associated with quasi-brittle materials, Figure 6. In addition, aggregate pull-out may lead to crack bridging, Figure 7. In matrix supported fabrics, the cracks have to propagate through the matrix for complete material separation to occur. In the matrix, interfaces are observed between regions of high and low porosity. Cracks that form have little sensitivity to other cracks in close proximity. They will commonly remain in the high porosity region rather than crossing into the low porosity region where higher energy would be needed for propagation, Figure 8 (Sorensen and Horsewell, 2001).

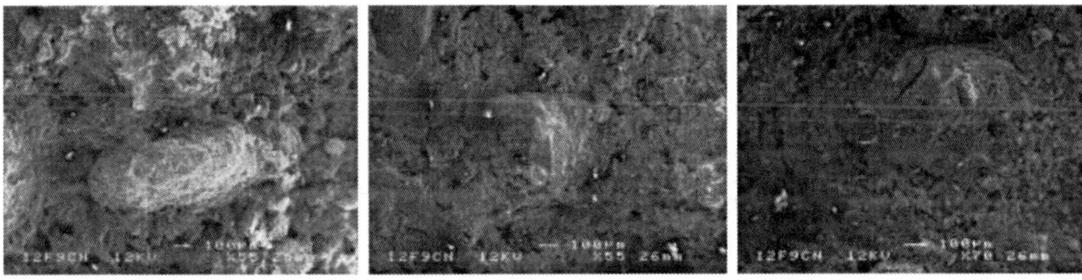

(A) (B) (C)

FIGURE 6: Evidence for crack propagation around the filler particles in un-irradiated Gilsocarbon graphite.

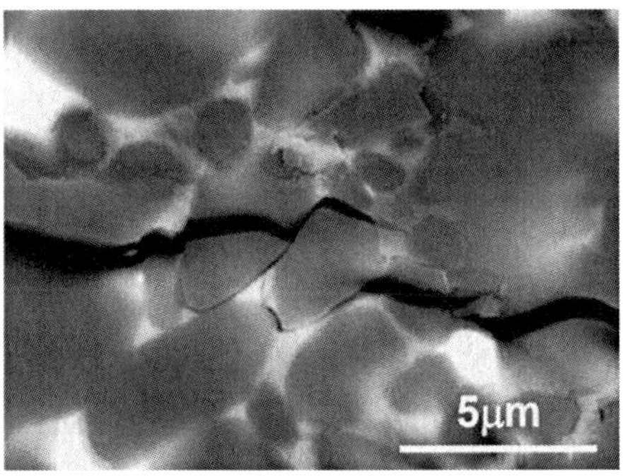

FIGURE 7: Image showing crack bridging by aggregates in SiC.

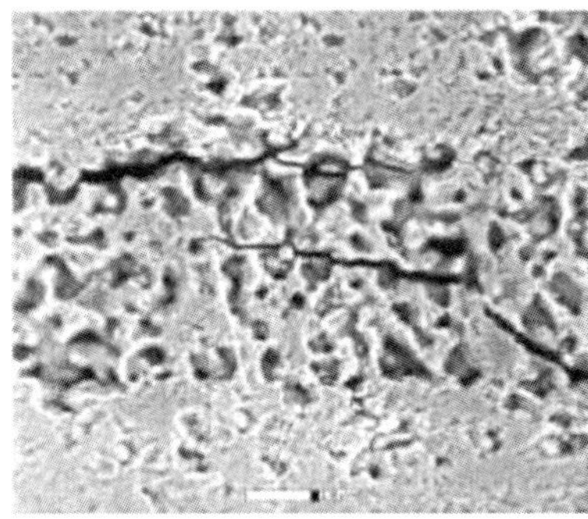

FIGURE 8: Image showing multiple cracks in close proximity in porous Al_2O_3.

Connectivity is maintained by distributed ligaments that are present within the skeletal structure that develop in materials with a high pore density. Under load, the ligaments behave like small beams and micro-cracking is commonly observed to progress by the brittle fracture of individual ligaments. As damage accumulates, strains may be accommodated, which limits strain energy storage. In regions of low pore density, more classical linear elastic crack-like features are commonly observed. However, both micro-crack initiation and propagation continues to be influenced by the microstructure. Small aggregates and crystallites act as obstacles causing cracks to deflect, bifurcate or arrest.

CONCLUSIONS

Quasi-brittle fracture behaviour is a complex interaction of different factors that produce different modes of failure under different load conditions. Random variations are observed in the measured values for the fracture properties. The variations are associated with the influence of variations in the microstructure and are only thought to be influential in small volumes. In addition, systematic variations in the fracture properties are associated with the influence of test specimen geometry, the size of the stressed volume and the loading mode (Becker, 2011). The systematic variations in the values obtained for the fracture properties may be qualitatively explained, in part, by reference to the distribution of the microstructure and microstructure sampling effects. In this paper, we have attempted to explain that differences in the way each of the microstructural phases behaves under load can influence damage accumulation and, by association, strain energy storage.

REFERENCES

Anderson, T. L. (1994). Fracture mechanics: fundamentals and applications, 2nd Edition, CRC Press, ISBN 978-0-8493-1656-2.

Bazant, Z. P. and Planas, J. (1998). Fracture and size effect in concrete and other quasibrittle materials. CRC Press, ISBN 0-8493-8284-X.

Becker, T. H. (2011). The double torsion beam technique: An experimental evaluation of the technique and its application to nuclear graphite. Thesis for the completion of a Masters in Mechanical Engineering University of Cape Town.

Boel, V., Audenaert, K. and DeSchutter, G. (2007). Characterisation of the microstructure of self-compacting concrete. *Proceedings of the 11th Euro-seminar on microscopy applied to building materials*.

Diamond, S. and Huang, J. (1998). Interfacial transition zone: reality or myth? *Proc. RILEM 2nd Int. Conference on the Interfacial Transition Zone in Cementitious Composites*, RILEM Publications, 35, 1–40.

Diamond, S. and Huang, J. (2001). The ITZ of concrete – a different view based on image analysis and SEM observations. *Cement and Concrete Composites*, 23, 179–188.

Ewalds, H. L. and Wanhill, R. J. H. (1989). Fracture mechanics. Edward Arnold Press, ISBN 0-7131-3515-8.

Hodgkins, A. (2006). Crack propagation in nuclear graphite. PhD Thesis, University of Manchester.

Hodgkins, A., Marrow, T. J., Wootton, M. R., Moskovic, R. and Flewitt, P. E. J. (2010). Fracture behaviour of radiolytically oxidised reactor core graphites: a view. *Materials Science and Technology*, 26, [8], 899 – 907.

Sorensen, B. F., Horsewell, A. (2001). Crack growth along interfaces in porous ceramic layers. *Journal of the American Ceramic Society*, 84, 2051 – 2059.

Tucker, M. E. (2001). Sedimentary petrology, an introduction to the origin of sedimentary Rocks. 3rd Ed, Blackwell Press, ISBN 0-632-05735-1.

Development of an Experimental and Simulation Process to Determine the End of Life Radionuclide Inventory of UK Irradiated Graphite Waste

Greg Black, Abbie N. Jones and Barry J. Marsden
Nuclear Graphite Research Group, School of MACE, University of Manchester, Manchester, UK, M13 9PL
Email: greg.black@postgrad.manchester.ac.uk

Abstract
When all operational reactors are shutdown, the UK will have a graphite waste legacy of 99,000 tonnes; this waste contains a variety of radionuclides and will fall under the Intermediate Level Waste classification for which the UK does not have a current disposal policy. Characterisation of the end of life radionuclide inventory is required before any policy can be developed. This paper investigates a combined programme of experimental and simulated work has been undertaken to investigate this. The tritium, carbon-14 and cobalt-60 content of a trepanned sample from Wylfa Magnox Reactor 1 have been experimentally determined using beta liquid scintillation counting and gamma spectroscopy. The WIMS9A reactor code and FISPACT-2007 neutron activation software have also been used to calculate this samples' inventory considering only a model which is isolated from the reactor circuit. Comparison between experimental and calculated results has shown that the original impurity levels are sufficient to explain the end of life activity, without additional consideration of contamination from other materials in the reactor circuit, in this type of simulation. Additionally, the calculations show that the production of ^{14}C from ^{14}N is approximately equal to that produced from ^{13}C.

Keywords
Decommissioning, waste, inventory

INTRODUCTION

The UK has predominantly used graphite moderated reactors for research purposes and both the production of electricity and weapons material. Consequently, the Nuclear Decommissioning Authority (NDA) has estimated when all reactors are shutdown this will represent a graphite waste legacy of 99,000 tonnes (NDA, 2002, NDA, 2007). Graphite wastes are known to contain a wide range of radionuclides, including ^{3}H, ^{14}C and ^{60}Co, which arise from both activation of impurities present in the graphite at start of life and from contamination from other components in the reactor circuit. The majority of this material will be classified as Intermediate Level Waste (ILW), a category which the UK does not currently have a disposal route for (NDA, 2010). The preferred option under consideration for ILW is disposal in a deep geological repository, however, graphite wastes would account for 33% of any repository volume (NDA, 2007). Therefore, innovative treatment options are under investigation which may provide methods to remove a significant proportion of this activity thus reducing the bulk of the material from ILW to Low Level Waste (LLW); considerably reducing the disposal cost and volume (Banford *et al.*, 2008). To inform both direct disposal and possible treatment options will require accurate predictions of the end of life activity of the waste. As it would be impractical to perform large scale characterisation of the graphite, it is the aim of this work to develop models to predict the inventory. These models are developed using both reactor physics and activation codes and the results compared against experimental analysis performed at the University of Manchester on samples which have been trepanned from reactor cores.

MODELLING METHODOLOGY

During the operational life of both Magnox and AGR reactors, samples are periodically trepanned from graphite bricks for core integrity studies; these samples may also be analysed for radionuclide content. This combined with accurate information regarding the fuel rating in the channel provides sufficient data for both the development and validation of radionuclide inventory calculations. Given that the lifetime behaviour of graphite components in a nuclear reactor is complex with both structural and compositional changes (IAEA, 2006), the development of the calculations in this study has been split into three stages:

- Stage 1: Calculations in this stage consider an isolated system, where the original impurities in the graphite are the only possible source of activation, there is no contamination or removal of material from the system.

- Stage 2: Calculations in this stage will take into account radiolytic oxidation of the graphite, representing loss of material from the system.

- Stage 3: Representing a fully dynamic system, where account is taken of possible contamination of the graphite from other material in the reactor circuit such as the fuel, clad, coolant and other structural components.

The work described in this study represents Stage 1 type calculations.

A trepanned sample of mass 4.35g has been obtained from the Wyfla 1 Magnox reactor and examined in the active facilities at the University of Manchester. The Wylfa Reactor 1 was commissioned in 1971, the core is constructed of Pile Grade A (PGA) graphite moderator bricks, with 6,156 vertical fuel channels containing 49,248 fuel elements. This particular sample was trepanned in April 2007 from fuel channel 1319, and had experienced an average lifetime fuel rating of 2.88MW/t. The ^{3}H and ^{14}C content of the sample were measured using liquid scintillation beta counting and the ^{60}Co content using a NaI(Tl) gamma spectrometer.

In order to use reactor physics codes to simulate the radionuclide inventory of irradiated graphite samples, information is required in four areas:

1. Elemental composition of the graphite before irradiation
2. Neutron flux experienced during lifetime
3. Reaction cross-sections for the elements present
4. Decay pathways of the resultant radionuclides

PGA graphite is manufactured from a petroleum coke and coal tar binder (Nightingale, 1962) and involves several high temperature baking (~800°C), graphitising (~2800°C) and purification stages which remove most of the impurities. However, it is inevitable that trace impurities remain, additionally compounds used in the manufacturing process lead to further contamination of the final product (Gerlach *et al.*, 2006, Takahashi *et al.*, 1999). Historic information regarding the original impurity content of PGA graphite is limited, current simulations use information from a study undertaken by 1984 using mass spectrometry of the ash residue of UK graphite grades (White *et al.*, 1984); the derived composition from this study is detailed in Table I.

Impurity element	PGA (ppm)	Impurity element	PGA (ppm)	Impurity element	PGA (ppm)
Li	0.05	Ti	3	Cd	0.04
Be	0.02	V	12	In	0.05
B	0.1	Cr	0.35	Sn	0.05
N	10	Mn	0.04	Ba	1.5
Na	1.0	Fe	10	Sm	0.04
Mg	0.1	Co	0.02	Eu	0.004
Al	1.0	Ni	1.0	Gd	0.005
Si	35	Zn	0.13	Dy	0.008
S	50	Sr	0.4	W	0.12
Cl	2.0	Mo	0.1	Pb	0.12
Ca	35	Ag	0.001	Bi	0.08

TABLE I: Impurity content of PGA graphite experimentally determined by White *et al.*, 1984.

The ANSWERS software code WIMS9A (Serco, 2011) was used to calculate the neutron flux (n/cm^2) in the graphite moderator. The model is a two-dimensional 'pin-cell' of a single fuel channel surrounded by a reflective boundary representing whole core conditions. The calculation involves a burn-up cycle to simulate the fuel performance at mid-life, an average of two years for Magnox fuel. The neutron flux averaged over the entire graphite moderator brick was output using 172 energy group structure which is shown in Figure 1.

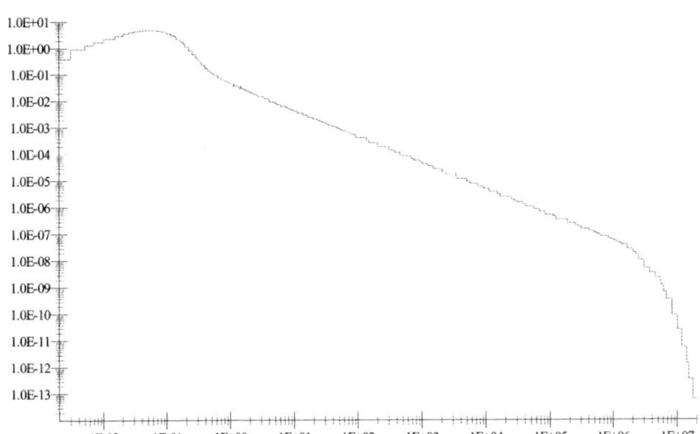

FIGURE 1: Normalized neutron flux in graphite in 172 energy groups for Magnox fuel at mid-life, calculated using the WIMS9A code.

The inventory calculation was performed using the United Kingdom Atomic Energy Authority (UKAEA) neutron activation software FISPACT (Forrest and EURATOM / UKAEA, 2007). This software uses the 2007 version of the European Activation Files (EAF), which contains an extensive data library of 65,565 neutron induced cross-sections and a database of decay information for 2,231 nuclides (Forrest and EURATOM/UKAEA, 2007). This software is used to determine the result of exposure of a group of elements to a neutron flux, integrating this over a specified time period and calculating the final activity and production pathways. These simulations outlined the sample to have a mass 4.35g and density 1.74g/cm^3 with an elemental composition as detailed in Table I. In addition, the operational and shutdown times of the reactor are included in the calculation; this is particularly important if a precursor nuclide has a short decay time whereby it would decay during the shutdown period before it can undergo further neutron activation. For this reason, the FISPACT simulation can incorporate shutdown times for the reactor by setting the flux at these time intervals to zero.

DISCUSSION

The experimental and simulated results are compared in Table II and show that the calculated results are greater than the experimental values in all cases. The calculated activities ranged from 36% for ^{14}C, 24% for ^{60}Co and 120% for ^{3}H, which, considering the uncertainties, are reasonably close for this type of comparison. The predictions also show that the quantities of the impurity elements assumed to be present, based on historic data, at start of life are sufficient to account for the experimentally determined end-of-life activity of the sample before including any additional contamination which may come from the reactor circuit.

The primary source of uncertainty in the results is recognised to arise from the impurity composition. The impurity levels can have a significant effect on the quantity of radionuclide produced; in addition, for tritium, it is reasonable to assume that, given the high mobility, a considerable fraction of the original activity may be lost before, or during experimental analysis (Marsden and Wickham, 1998, NDA, 2007, Wickham *et al.*, 2006).

	Experimental		Simulation	
Radionuclide	Activity (Bq/g)	uncertainty (%)	Activity (Bq/g)	C/E
^{3}H	2.08E5	10	4.57E5	2.20
^{14}C	5.09E4	10	6.93E4	1.36
^{60}Co	2.37E4	5	2.93E4	1.24

TABLE II: Comparison of experimental and simulated results for ^{3}H, ^{14}C and ^{60}Co, for Wylfa sample 1319/12.

The calculated production pathways from FISPACT are given in Table III and illustrate that both tritium and ^{60}Co are produced from single precursor elements; ^{6}Li and ^{59}Co, respectively. ^{14}C is produced via ^{13}C and ^{14}N, with an approximately 50/50 split between these two routes. This latter finding is an important result as there is currently debate (von Lensa, 2008) about which of these reactions is the predominant route for ^{14}C production in irradiated graphite waste.

Radionuclide	Production Pathway	Relative contribution (%)
^{3}H	$^{6}Li\ (n,a)\ ^{3}H$	99.461
	$^{6}Li\ (n,a)\ ^{3}H\ (b-)\ ^{3}He(n,p)\ ^{3}H$	0.508
^{14}C	$^{14}N\ (n,p)\ ^{14}C$	48.522
	$^{13}C(n,\gamma)\ ^{14}C$	51.476
^{60}Co	$^{59}Co\ (n,\gamma)\ ^{60}Co$	42.483
	$^{59}Co\ (n,\gamma)\ ^{60m}Co\ (IT)\ ^{60}Co$	53.163

TABLE III: Relative percentage of production pathways from original elements in graphite to ^{3}H, ^{14}C and ^{60}Co.

The relative cross-sections for ^{14}C production reactions are given in Table IV; for a 1% concentration of ^{13}C the reaction cross-section is 0.0009 barns and 0.001% ^{14}N inclusion the cross-section is 1.8barns. When considering the relative ratio of the abundance of both species within this virgin nuclear graphite this is inversely proportional to the ratio of the cross-sections therefore an almost equal contribution per reaction pathway for the production of ^{14}C is expected. This result is regardless of any net gain or loss, as is the case for this isolated system under consideration.

Reaction	Abundance of isotope in natural element (%)	Cross-section of reaction (barns)
^{14}N (n,p) ^{14}C	99.63	1.8
^{13}C (n,γ) ^{14}C	1.07	0.0009

TABLE IV: Abundance and cross-section information for C14 production routes (Wickham *et al.*, 2006)

CONCLUSIONS

Predictions of the end of life activity of ^{3}H, ^{14}C and ^{60}Co for Magnox graphite have been performed and compared against experimental results obtained from a trepanned sample taken from a moderator brick in the Wylfa Reactor 1. These predictions show good agreement for ^{14}C and ^{60}Co. The differences between the experimental and calculated results may be due to the assumed original impurity levels for the virgin graphite which were obtained from historic data. In addition, these calculations only take into account the original impurity content, whereas the behaviour of graphite in the core may include both loss of material, via radiolytic oxidation, and gain of impurities via contamination from other material in the reactor circuit. Future plans for this research include using mass spectrometry analysis to determine the impurity content of UK graphite grades: PGA and Gilsocarbon, as well as extensive sensitivity studies on the input data. The complexity of the models will also be expanded to reflect the dynamic processes experienced by the graphite in the core environment.

Acknowledgements
This work forms part of an EngD doctoral project and is funded by the Health and Safety Executive and the CARBOWASTE project (FP7-211333).

References

Banford, A., Eccles, H., Graves, M., von Lensa, W. & Norris, S. (2008). CARBOWASTE: An integrated approach to irradiated graphite. NNL.

Forrest, R. & EURATOM/UKAEA (2007). *FISPACT-2007 User Manual*, UKAEA.

Gerlach, D. C., Cliff, J. B., Hurley, D. E., Reid, B. D., Little, W. W., Wickham, A. J. & Simmons, T. A. (2006). *Applied Surface Science,* 252.

IAEA (2006). Characterization, treatment and conditioning of radioactive graphite from decommissioning of nuclear reactors. IAEA-TECDOC-1521.

Marsden, B. J. & Wickham, A. J. (1998). Graphite disposal options - a comparison of the approaches proposed by UK and Russian reactor operators. *Published Engineering Publishing*, 145-153.

NDA (2002). Radioactive wastes in the UK: A summary of the 2001 inventory. *Report No. DEFRA/RAS/02/003 Nirex Report N/041.*

NDA (2007). Final Waste Issues Group Report.

NDA (2010). Radioactive wastes in the UK: A summary of the 2010 inventory.

Nightingale, R. E. (1962). *Nuclear Graphite,* New York, Academic Press.

SERCO (2011). *The ANSWERS Software Service* [Online]. Available: http://www.sercoassurance.com/answers/ [Accessed 24th August 2011].

Takahashi, R., Toyahara, M., Maruki, S., Ueda, H. & Yamamoto, T. (1999). Investigation of morphology and impurity of nuclear grade graphite and leaching mechanisms of carbon-14. Manchester.

von Lensa, W. (2008). Proceedings of the 4th international topical meeting on high-temperature reactor technology. *In:* HTR2008, 2008 Washington D.C. USA.

White, I. F., Smith, G. M., Saunders, L. J. & Kaye, C. J. (1984). Assessment of graphite management modes for graphite from reactor decommissioning. EEC Report EUR9232 (Contract DE-D-001-UK).

The Oxidation of A3-3 Matrix Material in a CO_2 Atmosphere in Support of a Nuclear Battery Type Reactor Design

Joel Turner, Abbie N. Jones, Anthony J. Wickham, Marc Schmidt and Tim J. Abram

The University of Manchester, Oxford Road, UK. M13 9PL

Email: joel.turner@postgrad.manchester.ac.uk

Abstract

A significant attraction of a small-scale self-contained nuclear reactor that requires limited human intervention is a reduction in the required infrastructure costs through on-site generation. The option of CO_2 coolant rather than helium within a VHTR-type design has been considered, as the use of CO_2 results in a higher potential for convective cooling, and lower costs. Within a CO_2 coolant environment, graphite components are subject to oxidation, both through radiolysis and potentially thermally, and for modern nuclear graphites alongside the graphitic matrix A3-3, this behaviour must be considered as part of the design process. This research presents the results from irradiation experiments within pressurised sealed quartz ampoules, which have been conducted at The Technical University of Delft, alongside thermal oxidation experiments and material characterisation at The University of Manchester. Characterisation work has been undertaken to understand changes to the microstructure in terms of pre- and post-oxidation. The techniques applied include laser con-focal microscopy and helium pycnometry, with SEM and X-ray tomography planned for future work. The effects of oxidation on A3-3 matrix are compared to two grades of nuclear graphite: NBG-18 and Gilsocarbon with the effects of oxidation on TRISO coated particles also planned for investigation.

Keywords

A3-3 Matrix, oxidation, characterisation

INTRODUCTION

The 'Uranium-Battery' nuclear reactor concept, a self-contained small-scale design, based on the Very High Temperature Reactor (VHTR), has considered the use of CO_2 coolant at the design phase, in order to avoid the high costs associated with the use of helium. A further advantage is for coolant to circulate via convection, rather than through forced mechanical pumping.

The introduction of CO_2 coolant raises questions when used with a graphite moderator as it is known to cause oxidation of graphitic material. This effect is well known, as the majority of UK reactors are CO_2 cooled and graphite moderated. First and second generation UK reactors are based on PGA graphite and Gilsocarbon, respectively, neither of which is expected to be used for future build. As such the oxidation behaviour of modern graphitic material to be used in new reactors with a CO_2 environment must be a considered.

As well as considering modern nuclear graphites, the VHTR design upon which the Uranium-Battery design is based calls for the use of Tri-Isotropic (TRISO) coated particle fuel, in order to safely accommodate the high temperatures required. Coated particles are used within fuel compacts or pebbles formed from semi-graphitic A3-3 material. CO_2 has been used to thermally oxidise A3-3 material in previous studies (Hoinkis, 1986, 1991) but the exact rates across a range of temperatures are not well established. The same problem exists for the resulting mechanical effects on the material and the effects of radiolytic oxidation. Beyond this, the exposure of TRISO fuel itself to an oxidising environment needs to be considered, as the outer layer of coating is also carbon and could be expected to be similarly reactive.

BACKGROUND

A3-3 matrix material is a "graphite-like" material, which houses TRISO particles during their use. Its production differs from nuclear graphite manufacture due to the presence of uranium fuel kernels. This requires temperature be kept below that required to graphitize carbon as this is higher than the fuel melting point. For the production of an A3-3 fuel compact, a mixture of synthetic and natural graphite is ground to powder and mixed with phenolic thermo-setting resin. This powder is used to over-coat the TRISO fuel. The resulting mixture is heat-treated and pressed into shape a number of times before a final press and heat treatment at 50 MPa and up to 1800°C. Within a CO_2-cooled graphite-moderated reactor, two possible oxidation processes exist, both of which must be considered: thermal oxidation and radiolytic oxidation.

Thermal oxidation (Kelly and Marsden, 2000) occurs because high temperatures within the reactor facilitate the reaction:

$$C + CO_2 \rightarrow 2CO \quad [1]$$

converting solid carbon within the graphite into gaseous carbon monoxide (CO) and thus reducing the mass of graphite present. Thermal oxidation within graphite occurs in three temperature-dependent regimes depending upon the limiting factor controlling the rate of reaction. Regime 1 concerns oxidation limited by the rate of chemical reaction between the gas and graphite and occurs at the lowest oxidation temperatures: here the graphite is oxidised homogeneously throughout. Regime 2 occurs when the rate is limited by the diffusion of the gas into the pore structure, and the rate of oxidation is dependent upon the distance from the graphite surface. Regime 3 is limited by the rate of mass transfer of gas to the sample, and oxidation is confined to the sample surface. The differences in oxidation rate between A3-3 and other nuclear graphite should only occur in Regime 1 and (less so) in Regime 2, as the differences are likely to be dominated by microstructural porosity effects.

Radiolytic oxidation (Best, 1985) results from the effects of ionising radiation upon CO_2 molecules producing active species that are able to react with solid carbon to produce CO. The rate of the reaction is determined by coolant pressure and ionising-radiation flux and is strongly modified by diffusion effects which are introduced by the complex pore structure of the graphite. As the produced species is short ranged (~4-5μm under AGR operating conditions) (Best, 1985), it follows that small pores are more reactive than larger pores, and so the pore size distribution must be considered alongside the total pore volume.

There is some debate about which active species is responsible for the rate of oxidation observed in these conditions. Best (1985) showed that CO_3^- ions are an acceptable candidate supported by the production of CO_3^- within drift tube experiments and by theoretical calculation (Leary, 2000). McGurk (2000) has cast doubt on this assertion showing the production of CO_3^- from thermalized electrons, which form the vast majority within a reactor core, appears to be negligible when likely resonances in the ion-molecule interactions are considered: McGurk therefore opined that other mechanisms and intermediaries (O ions, HO_2

radicals) should be considered although when ions are absent (*e.g.* UV photolysis), different oxidising species are produced in the CO_2.

Experiments to support the 'Uranium Battery' concept must also account for the unique environment that the materials will experience. The coolant system is designed to be sealed, with very limited coolant chemistry control available. As can be seen from the above reactions, operation would result in the production of large amounts of CO from both mechanisms, and the effects of this need to be carefully considered. If these are too great, it is likely that the benefits of CO_2 use would be outweighed by the disadvantages, and a helium coolant would need to be considered.

EXPERIMENTAL METHOD

Both thermal and radiolytic oxidation processes must be considered experimentally for each graphitic material which implies the use of two concurrent experiments covering each oxidation process. Each experiment will be supported by the use of similar characterisation techniques, which are discussed separately. Experimental samples are cylindrical and divided into 'wide' samples, with a 6 mm diameter and a height of 4 mm, and 'tall' samples which are 4 mm in diameter and 6 mm high.

Thermal Oxidation Experiments

Thermal oxidation experiments were conducted using simultaneous thermal analysis (STA, *i.e.* combined differential scanning calorimeter – DSC and thermo-gravimetric analyzer - TG) for accurate rate determination of single samples.

STA oxidation experiments have been conducted to examine the effects of temperature on oxidation rate, and CO_2 flow rate at high temperature (up to 1100°C). For rate determination experiments, tall samples were used to fit the STA geometry. Samples were heated to the required temperature at 20°C / min in a nitrogen flow of 40 ml / min. The sample was then held at temperature for 15 minutes before the introduction of CO_2 at a flow rate of 75 ml / min, selected following preliminary experiments at 1100°C utilizing 'wide' samples. The temperature was maintained during oxidation. After the required time, which was varied for each temperature in order to produce a measurable oxidation rate, the CO_2 flow was stopped and the temperature reduced to ambient (20°C). Results from STA experiments were corrected for the effects of changes in buoyancy of the sample holder by running an identical program without a sample present. The collected DSC data has not been analysed at this point.

Radiolytic Oxidation Experiments

Irradiation experiments were conducted at The Technical University of Delft in The Netherlands. A3-3 and NBG-18 tall-samples were placed within sealed quartz ampoules containing CO_2 at a pressure of 11bar absolute. In order to produce these sealed ampoules, an open ampoule was produced with a neck to attach to a 'Swagelok' valve connected to a system containing a pressure transducer, analytical grade CO_2 bottle provided by BOC and a two-stage vacuum pump. The ampoule is sealed at pressure as the lower half of the ampoule is placed within liquid nitrogen freezing the CO_2. This allows the glassblower to melt and seal the neck of the ampoule.

After sealing, the ampoule is placed within the HOR reactor at TU Delft at a location providing an ionizing dose of 900 Gy/s, determined by polyethylene discolouration. Samples have been irradiated for 1100-4600 MGy total dose (15-60 full power days). The ampoule is then placed within a similar system to that used to produce it; however it is contained within a sealed Teflon sleeve allowing the ampoule to be broken with no loss of gas. This allows the calculation of any change in pressure following irradiation as well as providing the opportunity to take irradiated gas samples.

Materials Characterisation

Several techniques were used in order to characterise the material microstructure and pore structure, and the oxidation induced changes. These include con-focal microscopy and helium pycnometry, the techniques of which are discussed in detail in the literature. Future planned characterisation includes the use of tomography and SEM to support con-focal work, and BET alongside pycnometry in order to further quantify the pore structure.

RESULTS

Virgin Characterisation Studies

Samples that have been used for oxidation experiments were categorised beforehand, in order to determine individual effects on each sample. This included the use of helium pycnometry in order to determine the variation and mean value of open porosity present in each graphite type as well as to measure oxidation-induced changes. A statistical summary of the results is presented in Table 1. This is useful as incomplete virgin characterisation information is available for samples used in early irradiation studies, and the statistical analysis can be used to extrapolate this information.

Type	Number of Samples	Mean OPV (cm^3/g)	Standard Deviation
Gilso (Wide)	42	0.04737	0.0103
A3-3 (Wide)	96	0.05101	0.0146
NBG-18 (Wide)	76	0.02553	0.0044
A3-3 (Tall)	30	0.05998	0.0112
NBG-18 (Tall)	24	0.03595	0.0070

TABLE 1: Statistical summary of open pore volume measurements based on helium pycnometry from three graphitic materials.

These values would be expected to be lower than those calculated through mercury porosimetry or tomography because helium pycnometry operates only slightly above atmospheric pressure (1.344 bar) and so will not penetrate the smallest pores.

Some differences between material types are expected, as samples have been machined from graphite taken from different production billets. In the case of A3-3 material, 'wide' samples and 'tall' samples were produced from two separate pebbles. The extent of the difference between NBG-18 values is surprising however, especially given the small variation within the sample sets.

A3-3 and Gilsocarbon OPV values are shown to be significantly higher than NBG-18 for these samples. This is not surprising despite both Gilsocarbon and NBG-18 showing large volumes of porosity under microscopy and A3-3 not, as it has previously been suggested that a large majority of A3-3 porosity is contained within very small pores such as the ribbon network for glassy carbons suggested by Perret and Rutland (1972) and shown to exist in A3-3 by Hoinkis (1991).

Thermal Oxidation Studies

To date, the rate of oxidation of A3-3 material in CO_2 has been studied up to 1100°C. The results are presented in Figure 1, which follow the expected exponentially increasing rate with temperature. At 1025°C, the rate is 0.281 mg/min. The values presented here will be compared with both NBG-18 and Gilsocarbon under similar conditions.

Radiolytic Oxidation Studies

Irradiation experiments of A3-3 material have recently been completed at TU Delft. As a reference, a single ampoule of NBG-18 was irradiated for 14 full power days giving a total dose of ~1100 MGy. On inspection, following removal from the reactor, deposits were clearly visible on the surface of both the samples, and the quartz of the ampoule (Figure 2a). It is believed that these deposits are a result of CO build up within the ampoule, potentially combining with moisture present within the samples to produce polymeric material. Similar effects have been seen in operating Magnox plants within the UK, in low dose, low temperature regions (Wickham, 1989).

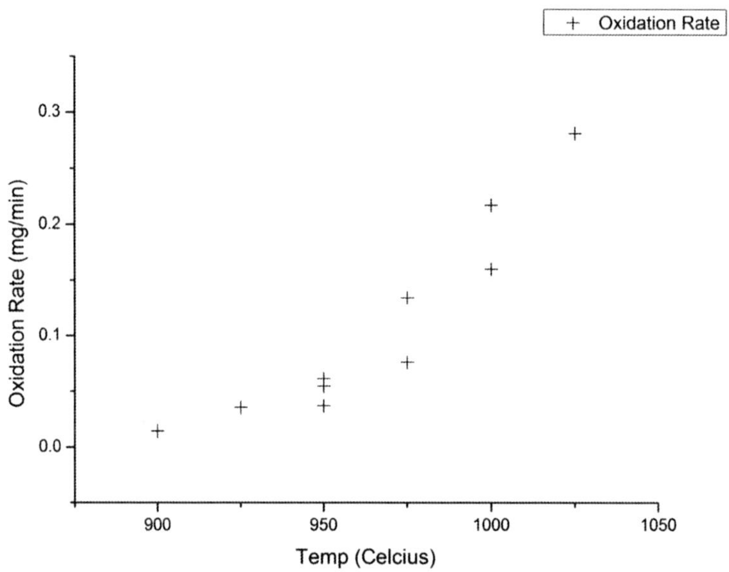

FIGURE 1: Effect of temperature on oxidation rate of A3-3 material at 75 ml/min CO_2 flow.

Following this result, A3-3 was irradiated for 14 days before being removed from the core, photographed and then returned to complete a 60 day irradiation programme. More significant deposits were seen on both the ampoule and the samples after 60 days (Figure 2b), although similar levels were observed after 14 days.

It is believed that the deposition of material on the sample surface would become an operational issue in a commercial plant with a sealed coolant system given the relatively short irradiation duration before the process was observed. In the UK Magnox and AGR plants, this problem has been overcome using a recombination unit consisting of a catalyst and oxygen supply. The use of such a unit would require forced circulation, however, removing the primary advantage that CO_2 provides in the U-Battery situation.

Characterisation of Radiolytically Oxidised Samples

Work has begun to characterise the radiolytically oxidised NBG-18 samples, which were the first to be irradiated. This has consisted of laser con-focal microscopy of the polished sample surface, after removal of surface deposits. To date, two surfaces from two samples have been imaged, Figure 3, in order to view porosity within the sample. Due to the nature of the polishing required for the technique, it is not possible to image the same surface before and after irradiation, and so a selection of images will be produced and compared to similar un-irradiated samples.

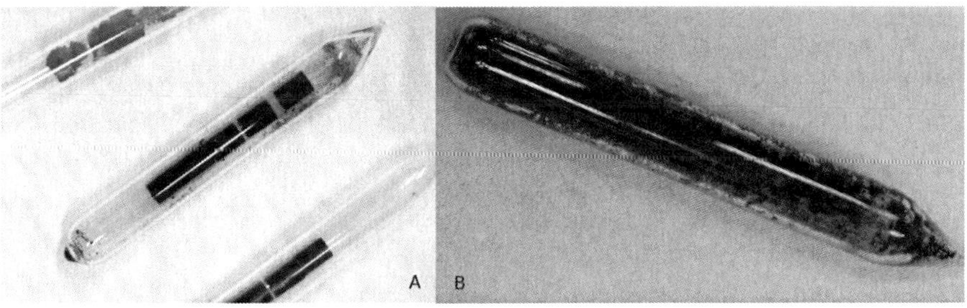

FIGURE 2: 'A' - NBG-18 Samples within an ampoule after 14 days irradiation (~1100 MGy) (center). Deposition is clearly visible on the surface of the samples and the ampoule itself. The upper ampoule visible contains polyethylene used for dose determination (after one hour exposure) and the lower ampoule is un-irradiated graphite. 'B' - A3-3 Material after 60 days irradiation time, showing large deposition of black material on the ampoule surface. Here the ampoule was more densely coated than the samples within.

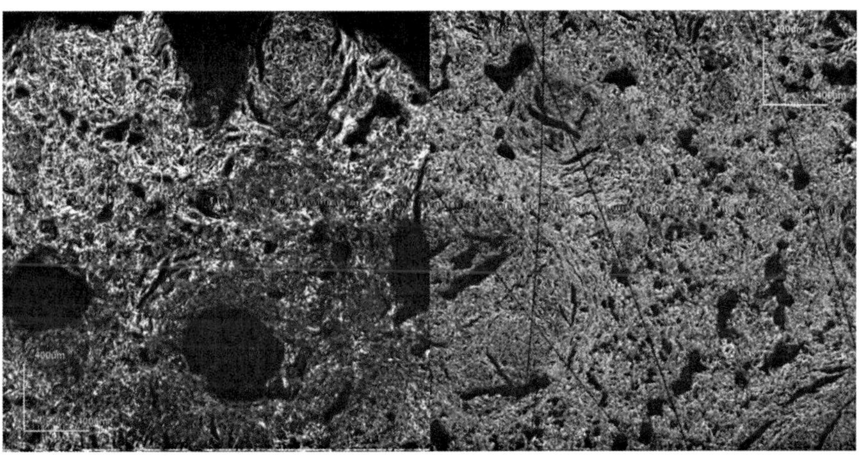

FIGURE 3: Images from 'Delft-1-1' and 'Delft-1-2' radiolytically oxidised NBG-18. Left: 5x optical image of Delft-1-1 showing large open porosity as well as a large volume of cracks between major pores. Right: 5x optical image of Delft-1-2, filler particles are clearly visible as is significant crack like porosity.

CONCLUSIONS

A study into the rate and behaviour of oxidation of A3-3 material, caused both thermally and through ionizing radiation, is well underway. Preliminary results suggest a significant increase in the rate of thermal oxidation at ~1000°C, with a flow rate of 75 ml/min CO_2. At 1025°C, the average rate is 0.281 mg/min.

Radiolytic oxidation studies show significant deposition on the sample surface and quartz container following small periods of irradiation, likely caused by a build up of reaction products within the sealed ampoule. These results are strongly suggestive of future difficulties when using an oxidising coolant within a sealed system and imply that any design utilizing a small-sealed CO_2 coolant loop will require monitoring of the coolant chemistry and the introduction of oxidation inhibitors in order to prevent excessive build up of reaction products.

Initial characterisation of radiolytically oxidised NBG-18 suggests the presence of large pores within the samples after a relatively short irradiation time. Further optical study of virgin material is necessary to properly determine if this is due to already present porosity or created via oxidation.

Acknowledgements
The authors would like to thank Urenco LTD for their support through the 'Uranium Battery' Project and the NWDA through their funding for the CNET research group.

References

Best, J. V., Stephen, W. J. and Wickham, A. J. (1985). Radiolytic graphite oxidation. *Progress in Nuclear Energy*, 16, [2], 127-178.

Kelly, B., & Marsden, B. (2000). *Irradiation damage in graphite due to fast neutrons in fission and fusion systems*. (Appendix 2 – Thermal Oxidation of Graphite) IAEA-TecDoc-1154.

Hinssen, H., Kuhn, K., & Moormann, R. (2008). Oxidation experiments and theoretical examinations on graphite materials relevant for the PBMR. *Nuclear Engineering and Design*, 238 (11) 3018–3025.

Hoinkis, E., Eatherly, W., & Krautwasser, P. (1986). Corrosion-and irradiation-induced porosity changes of a nuclear graphitic material. *Journal of Nuclear Materials,* 141-143, (Part 1), 87-95.

Hoinkis, E., & Allen, A. (1991). A study of precursory, original, and oxidized graphitic matrix A3-3 by small angle neutron scattering. *Carbon*, 29, (1), 93-100

Leary, P., Ewels, C., & Heggie, M. (2000). Modelling carbon for industry: radiolytic oxidation - computer simulation of materials at atomic level - Leary - Wiley Online Library.

McGurk, J. C. (2000). Formation of $CO_3^{=}$ in the Radiation Chemistry of CO_2. British Energy internal publication AEAT/R/NS/0042.

Perret, R., & Ruland, W. (1972). X-ray small-angle scattering of glassy carbon. *Journal of Applied Crystallography*, 5(3), 183–187. doi:10.1107/S0021889872009161

Wickham A.J.: seminar 'The Review of Safety at Magnox Nuclear Installations' – IMechE, London, 1989, 79-86

Part B

Empirical

Theoretical Investigations into Sample Size Effects on Ultrasonic Measurements of Elastic Moduli

John F. B. Payne and Nassia Tzelepi
National Nuclear Laboratory, Central Laboratory, Sellafield, Seascale, UK. CA20 1PG
Email: nassia.tzelepi@nnl.co.uk

Abstract
The relation between ultrasonic velocity and Young's modulus depends on sample dimension transverse to the direction of ultrasound propagation. For a sufficiently small transverse dimension, the phase and group velocities are the same and are independent of transverse dimension and independent of Poisson's ratio. For infinitely large transverse dimensions, the phase and group velocities are the same and dependent on Poisson's ratio. At intermediate sizes, several modes of ultrasonic wave propagation combine; the phase velocity is sensitive to transverse sample dimension, due to resonance effects and is not equal to the group velocity. The usual determination of ultrasonic velocity from the measured transit time of an ultrasonic pulse yields the group velocity. Analytical solutions consider the propagation of pure sine waves and consequently yield the phase velocity. This work extends previous analyses by considering propagation of a relatively short ultrasonic pulse, typical of those used for measurement of wave speed from the pulse propagation time. The main conclusion is that the phase velocity and transverse variation of the sine waves that do propagate are frequency dependent, resulting in distortion of a propagated pulse that is composed of sine waves of many frequencies. In the case studied here, the distortion is severe enough that the apparent wavespeed depends considerably on sample length and on the area of the sample surface in contact with the transducer.

Keywords
Ultrasonics, Young's modulus, size effects

INTRODUCTION

Young's modulus is a key parameter in the evaluation of internal stresses in graphite. Measurements on reactor material are used both to compile a database relevant to operating plant and to validate MTR data used for forward predictions. It is therefore important that these databases are supported by accurate reproducible measurements representative of the bulk material. Although there has been much experimental work to determine size effects and identify optimum experimental setup parameters and signal analysis (Tzelepi and Swallow, 2001; Tzelepi *et al.,* 2004 and 2005), there has been relatively little work on the theoretical understanding of these effects and their interdependence on apparent Young's modulus, calculated from the measured speed of propagation of an ultrasonic pulse. Most importantly, the experimental work has been carried out on unirradiated graphite and other materials, in which wave speeds differ from irradiated graphite. Therefore the conclusions drawn from the experimental work do not necessarily hold for measurements on irradiated graphite.

Graphite core monitoring campaigns currently use the ultrasonic time-of-flight (tof) technique to measure the Dynamic Young's Modulus (DYM). In the case of the Magnox reactors, the samples used are Ø12 × 6 mm (nominal) discs. The ultrasonic pulse speed through the thickness of the disk is measured using transmitting and receiving ultrasonic longitudinal transducers on opposite faces of the disk. This method yields the group velocity and its relation to the DYM depends on sample dimension transverse to the direction of ultrasound propagation. This is because the longitudinal wave speed depends on both the compression and shear resistance properties of the material. In a dispersive material, sound speed is a function of the sound frequency. Each frequency component propagates at its own phase velocity, while the energy of the disturbance or the modulation propagates at the group velocity. For sufficiently small transverse dimension, the phase and group velocities are the same and are independent of transverse dimension and independent of Poisson's ratio. For infinitely large transverse dimensions, the phase and group velocities are the same and dependent on Poisson's ratio.

At intermediate sizes, several modes of ultrasonic wave propagation combine; the phase velocity is sensitive to transverse sample dimension, due to resonance effects and is not equal to the group velocity. While the usual determination of ultrasonic velocity from the measured transit time of an ultrasonic pulse yields the group velocity, analytical solutions consider the propagation of pure sine waves and consequently yield the phase velocity.

This paper uses analytical and numerical methods to investigate wave propagation in a material of finite thickness. Specifically, waves are assumed to propagate in the z-direction and the material is assumed to be infinite in the y-direction and have finite thickness, *h*, in the x-direction.

ANALYTICAL METHOD

Wave propagation in cylinders was analysed by Love (1927) and wave propagation in a layer is discussed by Achenbach (1976). The analysis given here leads to the Rayleigh-Lamb frequency equation as given by Achenbach, but the interpretation of the significance of this equation differs from this.

Feynman gives the equation of motion in an elastic solid, as:

$$\rho \frac{\partial^2 \vec{u}}{\partial t^2} = (\lambda + \mu)\nabla(\nabla \cdot \vec{u}) + \mu \nabla^2 \vec{u} \qquad [1]$$

where $\vec{u}$ is the vector of elastic displacement and λ and μ are the Lamé elastic constants.

Boundary conditions at the free surfaces of a layer are such that, in general, a wave propagating along the layer has both shear and dilatation (compression) components. However, the different speeds of shear and dilatation waves produce different transverse variations for the same longitudinal variation and frequency, so it is simpler to separate the shear and dilatation components. Considering boundary conditions of zero tensile and shear stresses at the surfaces of the layer ($x = \pm h$) and ignoring flexural waves as unlikely to be excited in the particular experimental set-up, it can be shown that a wave whose z- and time variation is $\exp[i\,(\omega t - kz)]$ can be propagated, provided that:

$$\begin{aligned}\Phi(\xi,\Omega,\kappa) = 4\sqrt{\Omega^2-\xi^2}\sqrt{\kappa^2\Omega^2-\xi^2}\cos\left(\frac{\pi}{2}\sqrt{\kappa^2\Omega^2-\xi^2}\right)\sin\left(\frac{\pi}{2}\sqrt{\Omega^2-\xi^2}\right)\xi^2 + \\ (\kappa^2\Omega^2-2\xi^2)^2\cos\left(\frac{\pi}{2}\sqrt{\Omega^2-\xi^2}\right)\sin\left(\frac{\pi}{2}\sqrt{\kappa^2\Omega^2-\xi^2}\right) = 0\end{aligned} \qquad [2]$$

where $\xi = 2\,h\,k\,/\,\pi$ is the dimensionless wavenumber (*i.e.* angular wavenumber, *k*, is divided by the layer thickness), $\Omega = 2\,h\,\omega\,/\,(\pi\,c_D)$ is the dimensionless frequency (*i.e.* angular frequency, ω, is divided by the time needed for a dilatation wave to cross the layer thickness) and, $\kappa = c_D \ / c_S$ is the ratio of the phase velocity of the dilatation wave to that of the shear wave. This is the Rayleigh-Lamb frequency equation; it determines the possible wavenumbers, k, that correspond to a given frequency, ω and hence, the possible phase velocities ω/k, that correspond to a given frequency.

In Figure 1, Arg[Φ] is plotted as a function of ξ and Ω, for a Poisson's ratio of 0.24. The boundary between grey and white or between dark grey and black correspond to the trivial solution $\kappa\Omega = \xi$. The remaining boundaries between black and white and between dark grey and grey correspond to non-trivial solutions. The grey line corresponds to $\Omega = \xi$ and a phase velocity of c_D, the dilatation wave phase velocity.

Clearly there are many solutions to the frequency equation, which presents a difficulty in deciding which solutions are relevant. However, the phase velocities do not directly determine the relevance of the different waves as the speed of the transmitted signal is determined by the group velocity. If the spread of frequencies in the signal is not too large, the group velocity is $d\omega/dk$ and in terms of Ω and ξ:

$$\frac{d\omega}{dk} = \frac{d}{dk}\left(\frac{\Omega\pi c_D}{2h}\right) = \frac{d}{d\xi}\left(\frac{\Omega\pi c_D}{2h}\right)\frac{d\xi}{dk} = c_D\frac{d\Omega}{d\xi} \quad [3]$$

Analysis of a typical ultrasonic pulse through a sample of PGA graphite shows that it can be approximated by at least 8 sine waves with frequencies ranging from 260 to 400Hz, which is clearly a large spread of frequencies.

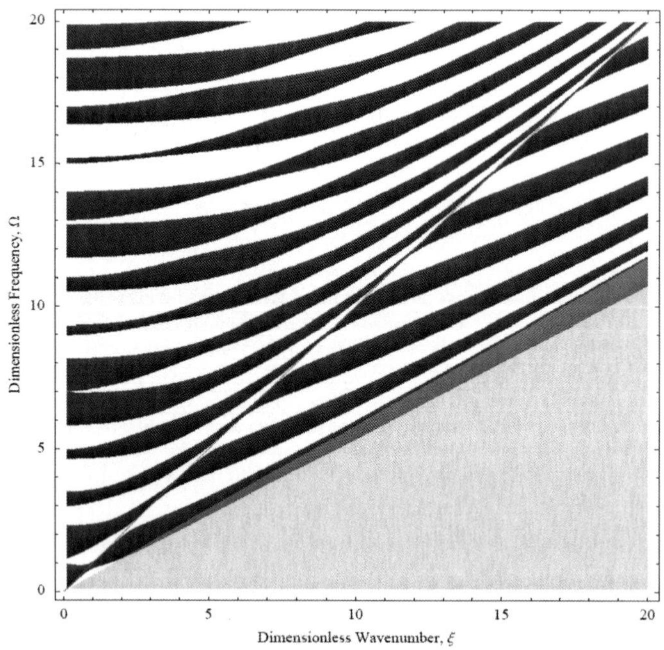

FIGURE 1: Plot of Arg[Φ(ξ,Ω,κ)] for a Poisson's ratio of 0.24.

FINITE ELEMENT ANALYSIS

The results reported here were produced by the finite element code "Impact". The "Impact" code is based on an explicit time-stepping algorithm and is therefore useful for time-dependent phenomena, such as wave propagation. "Impact" boundary conditions at nodes are set by specifying the velocity and acceleration at nodes. The y-accelerations and y-velocities at all nodes were set to zero in the boundary conditions. The boundary condition for all nodes at $z = 0$ set the z-velocities to vary in the same way as a typical ultrasonic pulse transmitted through PGA graphite as part of Young's modulus measurement (Payne, 2009). The sample extends from $x = -h$ to $x = +h$ and is symmetric about $x = 0$.

The drawback of solutions for wave propagation arrived at by numerical methods is that the results are not general: specific cases have to be chosen for solution. The results can be generalised to a certain extent by working with dimensionless variables; for example, results for a certain ultrasonic pulse frequency and sample size apply to any other ultrasonic pulse frequency and sample size that corresponds to the same dimensionless frequency and dimensionless wave speed ratio κ. However, a specific numerical result applies to only one dimensionless frequency. For this work, the specific cases were chosen with the intention that they would be the most interesting and illuminating cases. Therefore, a high value of Poisson's ratio, $\sigma = 0.4$ was chosen, because, at $\sigma = 0.4$, phase velocities differ more from the expected value for a large sample, c_D, than they do at lower Poisson's ratios.

Figure 2 shows Arg[Φ] plotted as a function of x and Ω, for Poisson's ratio of 0.4. The boundaries between different values of Arg[Φ] occur where $\Phi = 0$, the Φ value compatible with wave propagation. The boundary between grey (Arg[Φ] $= -\pi/2$) and white (Arg[Φ] $= \pi$) corresponds to the trivial solution $\kappa\,\Omega = \xi$. The remaining boundaries between black (Arg[Φ] $= 0$) and white correspond to non-trivial solutions. The diagonal grey line in Figure 2 corresponds to $\Omega = \xi$ and a phase velocity of c_D, the dilatation wave phase velocity. The other grey line in Figure 2 corresponds to the phase velocity expected from an analysis of wave propagation in a thin layer that assumes that tensile stresses in the x-direction and tensile strains in the y-direction are both zero.

From Figure 2, at a dimensionless frequency $\Omega \approx 0.75$ and $\sigma = 0.4$ there is only one non-trivial solution to $\Phi[\xi, \Omega, \kappa] = 0$ which is $\xi \approx 1.67$ corresponding to a phase velocity $\Omega/\xi \approx 0.449\ c_D$. Therefore, the sample dimensions and material properties were chosen such that the centre frequency of the ultrasonic pulse corresponded to $\Omega \approx 0.75$. Dimensionless wavenumber is defined as $\xi = 2hk\ /\ \pi$ and wavelength is $2\pi\ /\ k$ (since the wave is $\exp[i(\omega t - kz)]$), so the wavelength is $4h\ /\ \xi$. Hence, a dimensionless wavenumber $\xi = 1.67$ corresponds to a wavelength $\approx 2.4\ h$. The material properties were chosen to be: $Y = 7.5$GPa, $\rho = 1.2$g/cm^3 and $\sigma = 0.4$, which implies $c_D \approx 3.66$ mm/μs.

No boundary condition was imposed at $x = h$, consequently this sample boundary was free to move. The centre frequency of the pulse applied at $z = 0$ was 340 kHz which corresponds to a dimensionless frequency $\Omega \approx 0.743$.

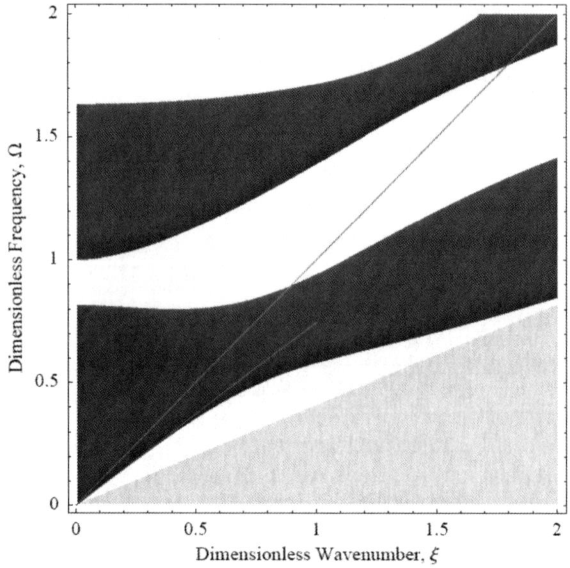

FIGURE 2: Plot of Arg[Φ(ξ,Ω,κ) for a Poisson's ratio of 0.4.

"IMPACT" FE RESULTS AND DISCUSSION

As expected with $\Omega \approx 0.75$, the behaviour of the propagated pulse is not simple. Figure 3 shows the wave propagating in the sample: displacement in the z-direction is plotted against z, the direction of wave propagation. At 6.5 μs from the start of the pulse, the z-displacements at the centre and the edge of the sample ($x = 0$ and $x = \pm h$, respectively) are plotted in black and grey, respectively. At 4.3μs from the start of the pulse, the z-displacements at the centre and the edge of the sample are plotted in black and grey, respectively. The z-displacements at the centre of the sample at 4.3 μs (black points) are offset by 6.3 mm and scaled by a factor 0.77 so as to align the first trough with the z-displacements at the centre of the sample at 6.5 μs (black points). The z-displacements at the edge of the sample at 4.3 μs (grey points) are offset by 6.6 mm and scaled by a factor 0.77 so as to align the first trough and first peak with the z-displacements at the edge of the sample at 6.5μs (grey points).

Clearly, the wave shape at the centre of the sample ($x = 0$) changes considerably with time. To align the first peaks of the pulses, an offset of 5.0 mm is required, which is quite different from the offset required to align the troughs. Furthermore, although, at the edge of the sample, a single offset aligns both the first peak and the first trough, the relative magnitudes of the peak and trough at 4.3μs differ from their relative magnitudes at 6.5 μs.

From Figure 3, the ultrasonic pulse has travelled between ~5.0 mm and ~6.6 mm in between 4.3 μs and 6.5 μs (depending whether the pulse is timed at the edge or middle of the sample) so the apparent speed of propagation of the pulse is ~ 2.3 to 3.0 mm/μs. A speed of 2.3 mm/μs is indeed much less than c_D, the speed expected in an infinite medium, which is ~3.66 mm/μs, although not as low as the phase velocity at the pulse centre frequency ~$0.449c_D$.

As discussed by Payne (2002 and 2009), only waves with specific transverse variation of displacements propagate. Wave components with any other transverse variation are attenuated exponentially. This is illustrated in Figure 4 which shows the z-displacements in the material at 6.5 μs from the start of the pulse. The transverse (x-direction) variation of the z-displacement of the ultrasonic pulse changes as it propagates into the material. At $z = 0$, the boundary condition makes the z-displacement independent of x. This x-variation has components that correspond to waves that cannot propagate, so the dependence of z-displacement on x in the propagating wave varies strongly with x, even changing sign, as is obvious at $z = 8$ mm in Figure 4. To obtain a wave that propagates without change of shape, and therefore yields a meaningful group velocity, it is necessary to choose the x-dependence of z-displacement at $x = 0$ to correspond entirely to waves that can propagate.

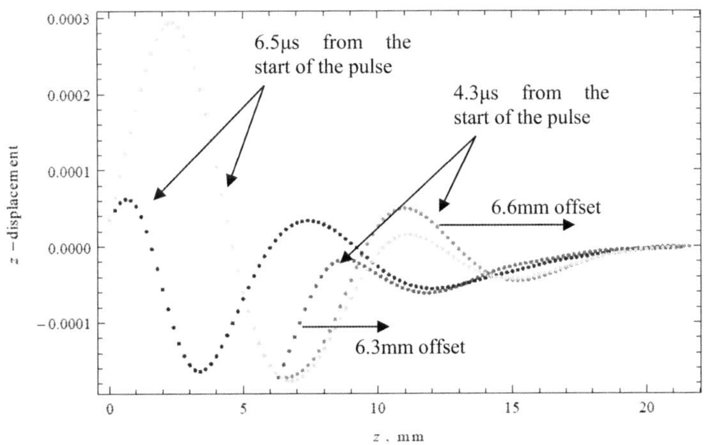

FIGURE 3: Variation of z-displacement with z.

Even if this is done, the variation of phase velocity of the pure sine wave with frequency may result in some distortion of the wave shape (z-displacement as a function of z) as the wave propagates. Furthermore, the variation of the propagatable x-dependence of z-displacement with frequency of the pure sine wave, together with the variation of phase velocity of the pure sine wave with frequency may result in some distortion of another aspect of wave shape (z-displacement as a function of x) as the wave propagates.

In practical measurements (as opposed to numerical calculations) it would be difficult to control the x-dependence of z-displacement at the sample surface. This depends on the transducer design and the coupling between the transducer and the material. The x variation might be made to correspond more nearly to that of a wave with z-displacements of opposite sign at $x = 0$ and $x = \pm h$ by using a transducer smaller than the sample end face.

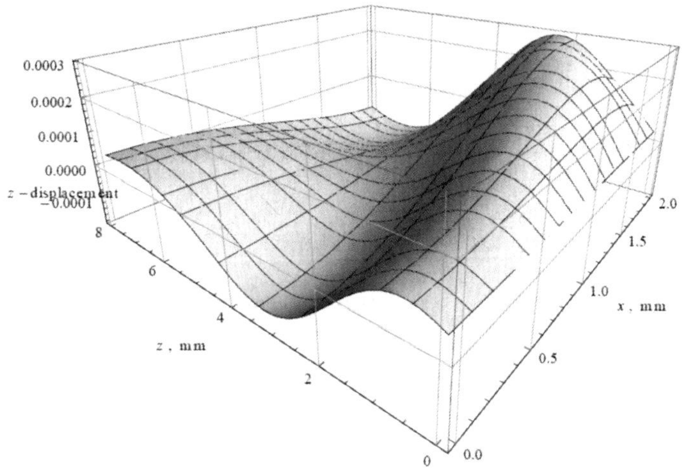

FIGURE 4: Variation of z-displacement with x and z.

In practice, also, the variation of *z*-displacement with *z* is difficult to measure, and instead the variation of *z*-displacement with time at the sample surface is measured. The dependence of *z*-displacement on *x* changes with time, for the same reason that it changes with *z*. Also the dependence of *z*-displacement on time changes as the wave propagates. In the case considered, the distortion is particularly severe at the centre of the sample ($x = 0$) and near the beginning of the pulse. Therefore the apparent speed of propagation changes, depending on the *x*-position at which the wave is received, and the point on the waveform used for timing. In this case, the apparent speed of propagation is between ~2.0 and 3.3 mm/μs. Variation in contact conditions between transducer and sample across the end-face would change the contributions to the transducer signal from different *x*-positions which could change the measured speed at which a pulse propagated through the sample.

If the wave is received by a transducer that covers the end of the sample, and the transducer is equally sensitive at all points, then the received waveform will be the average of the *z*-displacements from $x = 0$ to $x = h$. Figure 5 shows the average *z*-displacements at $z = 0$ (black points), $z = 6$mm (grey points) and $z = 12$mm (dark grey points). The *z*-displacements of the $z = 6$ mm points are scaled by a factor 1.4 and the times are offset by ~ -2.1 μs in order to align the first trough and peak of the wave received at $z = 6$ mm with the wave transmitted at $z = 0$. The *z*-displacements of the $z = 12$ mm points are scaled by a factor 2 and the times are offset by -4 μs in order to align the first peak of the wave received at $z = 12$ mm with the wave transmitted at $z = 0$. Other choices of alignment points would require different time offsets. The shape of the wave changes as it propagates, so the apparent wave speed depends on the point on the waveform that is used for timing, and on the sample thickness. The shape change is particularly obvious at $z = 12$ mm, where the *z*-displacement remains negative at the first peak, so a zero crossing between the first trough and peak cannot be used to time the pulse, because there is no such zero crossing. Aligning on the first peak, the apparent speed of propagation based on *x*-averaged *z*-displacement is between ~ 2.8 and 3 mm/μs.

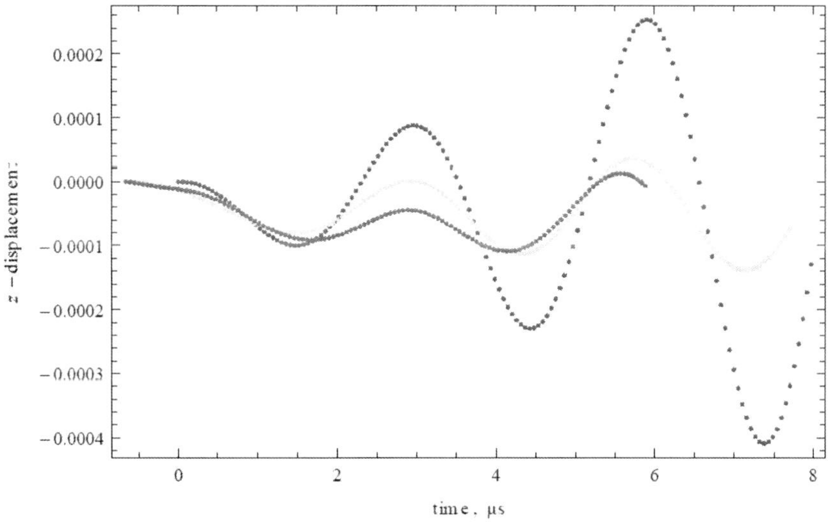

FIGURE 5: Variation of x-averaged z-displacement with time.

A further significant effect is not included in the calculations resulting in Figure 5. In practice, the acoustic impedance of the receiving transducer and the graphite sample are unlikely to match so reflections will occur at the sample-transducer interface. Such reflections have been excluded from the calculations by making the sample much longer than 6 mm.

CONCLUSIONS

Young's modulus is a key parameter in the evaluation of internal stresses in graphite. Measurements on reactor material are used both to compile a database relevant to operating plant and to validate MTR data used for forward predictions. It is therefore important that these databases are supported by accurate reproducible measurements representative of the bulk material. Although there has been much experimental work to determine size effects and identify optimum experimental setup parameters and signal analysis, there has been relatively little work on the theoretical understanding of these effects and their interdependence on apparent Young's modulus calculated from the measured speed of propagation of an ultrasonic pulse. Most importantly, the experimental work has been carried out on unirradiated graphite and other materials in which wave speeds differ from irradiated graphite. Therefore, the conclusions drawn from the experimental work do not necessarily hold for measurements on irradiated graphite.

This study used analytical and numerical methods and has shown that:

1. For a given frequency, there are many solutions to the frequency equation, which yields the phase velocity. However, the phase velocities do not directly determine the relevance of the different waves as the speed of the transmitted signal is determined by the group velocity. Although the latter can easily be calculated when the spread of frequencies in the signal is not too large, an ultrasonic pulse that is useful for timing purposes must have a well-defined start and a short duration and consequently a wide spread of frequencies.

2. In the chosen case, the dispersion and the failure to propagate of certain components of the wave cause considerable distortion of the propagated pulse. Frequency dependent attenuation of ultrasound in graphite was already known to distort the pulse shape; the effects found here are additional sources of pulse distortion. The pulse distortion is severe enough that the apparent wavespeed depends considerably on sample length and on whether the centre or edge of the sample contributes most to the electrical signal produced by the receiving transducer.

3. The distortion of the propagated pulse may cause a transducer that covers the entire end of the sample to produce a relatively small signal because displacements at the sample edge and centre produce opposite effects in the transducer.

References

Achenbach, J. D. (1976). Wave propagation in elastic solids, North Holland Publications.

Love, A. E. H. (1927). A treatise on the mathematical theory of elasticity. Fourth Edition, Dover Publications.

Payne, J. F. B. (2002). The speed of elastic waves in cylinders, BNFL R&T report.

Payne, J. F. B. (2009). The relation between elastic properties and wave speed in bodies of finite transverse dimension. National Nuclear Laboratories report.

Tzelepi, A. and Swallow, K. (2001). An investigation into the through transmission time-of-flight method for measuring the DYM of graphite, BNFL report.

Tzelepi, A., Swallow, K. and Lingham, I. (2004). Uncertainties in DYM measurements, BNFL report.

Tzelepi, A., Yates, G. and Lingham, I. (2005). Ultrasonic time-of-flight technique - DYM Mk3 proving trials, BNFL report

Mechanical Property Measurements on AGR Core Graphite Using Electronic Speckle Pattern Interferometry

Paul Ramsay
National Nuclear Laboratory, Sellafield, Cumbria, UK. CA20 1PF
Email: paul.ramsay@nnl.co.uk

Abstract
The continued strength and integrity of the graphite moderator in Advanced Gas-cooled Reactors (AGRs) is an important consideration for safe operation and plant life extension. Accordingly, samples of core graphite are routinely taken for monitoring of its physical properties, including the elastic modulus. Elastic modulus is currently measured using an ultrasonics technique that requires knowledge of the Poisson's ratio. However, although the Poisson's ratio of virgin graphite has been measured, there is little information available on the Poisson's ratio of irradiated (and oxidised) graphite. A new non-contacting technique has been developed, which uses Electronic Speckle Pattern Interferometry (ESPI) to accurately measure the 3-dimensional displacements which occur during a 3-point bend test. These displacements can be used to determine both Young's modulus and Poisson's ratio over the range of loading conditions up to failure. Development of the technique revealed that assumptions normally made during such mechanical testing were not necessarily valid when compared to the actual 3-D motions measured by ESPI. Extensive development was required both to understand and overcome these effects by comparing the measured movements with those predicted by 3-D finite element modelling. A new method is described here which can successfully measure both Young's modulus and Poisson's ratio using the small irradiated Gilsocarbon samples extracted from the AGR reactors.

Keywords
ESPI, Young's modulus, Poisson's ratio

INTRODUCTION

The physical properties of the graphite moderator in the AGRs are monitored by means of cylindrical samples extracted from the fuel channel bricks. Among the properties measured are strength, dynamic Young's modulus (DYM) and the coefficient of thermal expansion (CTE). The trepanned samples are up to 75 mm long and are 19 mm diameter. Since there is a requirement to measure properties related to the radial dose profile in the bricks, the cylinders are sliced into discs measuring 19 mm diameter by 6 mm thick. These are then subjected to DYM measurements using a time of flight ultrasonic method. Fracture strength measurements are made on small beams, measuring 6 mm x 6 mm x 19 mm, which are cut from the discs. It should be noted that a number of alternative techniques were considered (including Digital Image Correlation), but ESPI was identified as the best method for measurement of the 'static' modulus and Poisson's ratio.

IMPLEMENTATION

The implementation of ESPI was designed around an existing routine technique where irradiated Gilsocarbon is subjected to a 3-point bend test to failure. This enabled the measurement of the sample displacements and hence strains during the test which can then be used to calculate both Young's modulus and Poisson's ratio.

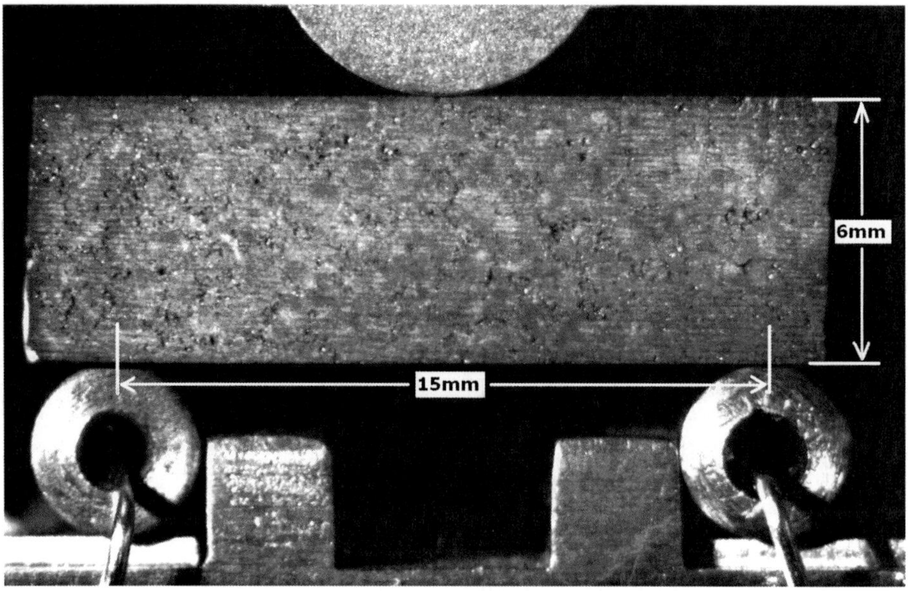

FIGURE 1: ESPI field of view of sample under test.

Figure 1 shows the view from the ESPI camera with a virgin graphite beam mounted on the test fixture. The particular requirements for ESPI and the small sample size presented a number of challenges:

- Since ESPI measures the difference in the optical path length between a pair of coherent light beams, any relative motion between the camera and sample due to vibration may cause the correlation between the beams to be lost. To limit the effect of vibration, the camera was supported directly from the test frame by heavy gauge aluminium plates. The whole test machine was also installed on air suspension mounts to limit the transfer of vibration from the enclosing fumehood.

- Any interruption of the beams will prevent a complete displacement map from being generated, therefore the design included rollers that did not protrude beyond the face of the sample thereby eliminating shadows.

- To prevent the sample twisting during the test, the loading and support rollers need to be parallel and in contact with the sample across its full breadth. The lower rollers articulate independently on 'D' shaped shoes, which adjust during pre-loading to accommodate small deficiencies in the squareness of the samples resulting from the remote machining operations within shielded facilities.

- The ESPI camera records images at regular time intervals during which an optimal amount of strain needs to occur. Too small a change will introduce significant noise since the total strain is the sum of all previously measured values. Too large a spacing will adversely limit the number of profile points. Using estimates of the expected failure load, based on a sample density, allows the test to be carried out at a constant load ramp rate which then gives a pre-determined number of profile points.

Methods

The dedicated ESPI computer system processes the image files that are recorded during a test and outputs data files containing the relative displacements between pairs of images. Three arrays contain the displacements occurring in each dimension for each image pixel. Figure 2 shows parts of the data after a typical load increment has been applied to the beam.

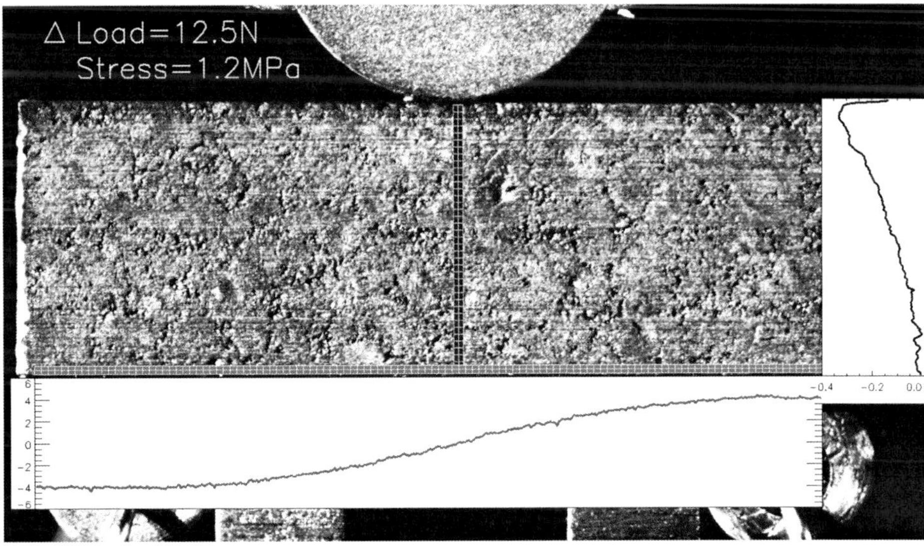

FIGURE 2: Example of ESPI output data for graphite sample under test.

The methodology required to determine the elastic properties from these data has developed over time. In particular, the development of 3-D finite element models that indicate how the sample should deform has proved to be of great assistance. The analysis route has developed through three distinct stages: direct method; line method; and area method.

Direct Method

The X dimension displacements along the bottom edge of the beam, lower graph in Figure 2, clearly show an overall length increase of around 8 μm. The X strain is easily determined from the gradient of this curve; the bending stresses from a concentrated load are determined from the applied load and beam dimensions (Timoshenko, 1970). The Y displacements down the centre line of the sample, right graph Figure 2, show a total of around 0.3 μm compression. Originally, it was expected that a linear fit to the lower portion of these data would be adequate to determine the Y strain.

Line Method

Following analysis of 3-D finite element models of a 3-point bend test, it was found that the Y displacement curve required a third order fit to extract the Y strain at the bottom surface. A comparison of the measured centreline response with that from computer models allowed for a determination of Poisson's ratio for homogeneous material, but when applied to graphite the observed 'granular' nature of the data prevented suitable fitting.

Area Method

By comparing the entire X displacement data to that from computer model outputs, it was possible to converge on values for both Young's modulus and Poisson's ratio. However, the Y displacement data was again found to be too noisy for accurate determination of Poisson's ratio, especially for irradiated graphite.

RESULTS

The stress-strain curve for a virgin graphite 3-point bend test sample is shown in Figure 3. The sample was load cycled three times to approximately 50% of its expected failure stress. During a later test the same sample was retested with the baseline load increasing during each cycle until failure, the complete loading profile is shown in the Figure 3 insert.

The expected stress-strain profile was observed, as shown in Figure 3, with the accumulation of permanent strain during unloading. Each load / unload cycle clearly shows some hysteresis and that the average Young's modulus during these cycles is constant irrespective of the absolute stress.

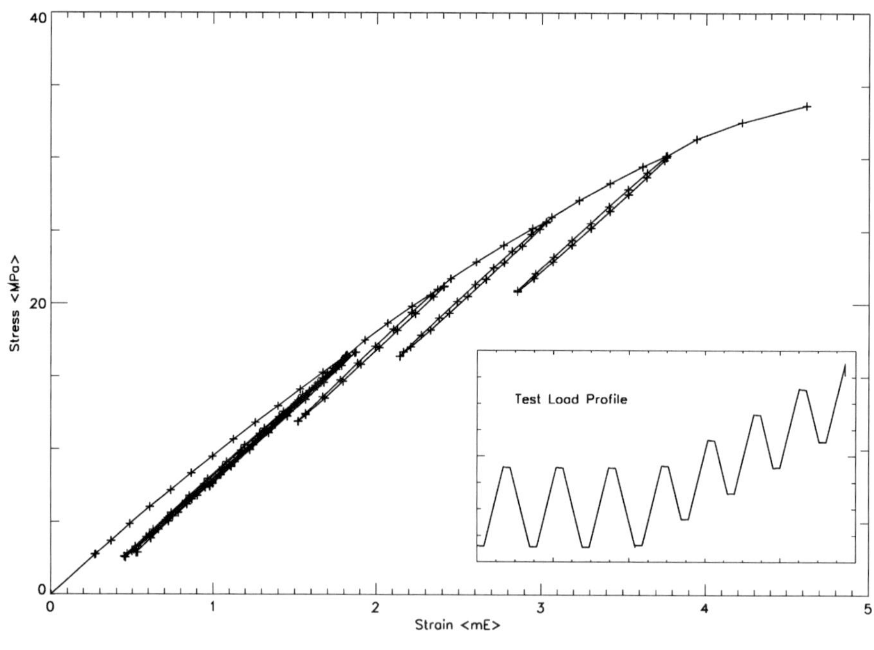

FIGURE 3: Stress-strain curve for virgin graphite sample.

A convenient way of displaying the results from ESPI is to show the individual values for modulus at each image point determined from the differential of the stress-strain curve. If the stress at that point is greater than any previously observed then the modulus is plotted against absolute stress, otherwise the modulus values are plotted against the stress level relative to that at the start of the current load ramp section. Figure 4 shows a graphite sample loaded three times to 50% failure stress and then loaded continuously until failure. When displayed this way it is clear that each trace extrapolates back to a single value at zero stress. This is equivalent to the DYM value at low strain that is measured by ultrasonic and resonance techniques.

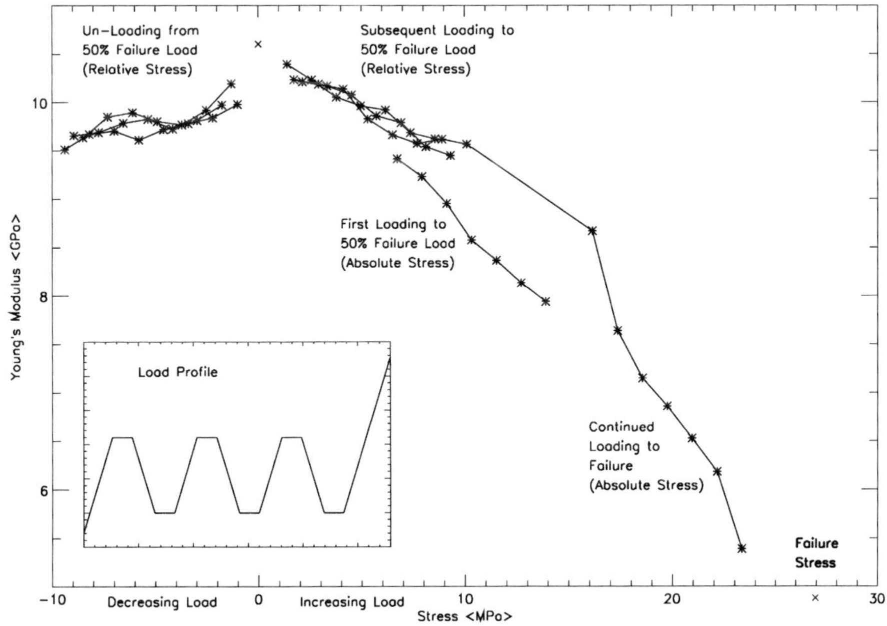

FIGURE 4: Response of virgin graphite beam sample to cyclic loading.

Unlike with Young's modulus, the results for Poisson's ratio on virgin graphite are always dependent on absolute stress value as Figure 5 shows. Values which occur after the permanent set has taken place show that Poisson's ratio continues to drop linearly as the absolute stress increases. This cannot be completely explained by the effect of micro-cracks within the graphite, as micro-cracks would be expected to progressively reduce the elastic modulus during intermediate unloadings and re-loadings. The application of a tensile stress is clearly causing more complex processes than purely micro-cracking to occur in the unirradiated graphite. Since there are no obvious visual changes in the condition of the graphite, these changes could be at much smaller crystallite and poly crystallite scales, but are currently not known.

Micro-cracks

During tests on graphite, examination of the X displacement traces, showed that more than half exhibited discontinuities where the material showed extreme local strain values. These effects can be globally visualised by plotting the total X displacement array as a surface plot viewed vertically and where the surface is shaded by an oblique light source. Figure 6 shows a virgin graphite sample close to failure. For this particular sample, the long crack on the centreline, which is located in a lower stress region was seen to grow towards the sample's bottom edge with increasing load. The optical image taken after failure confirmed that the failure crack did follow this path and was associated with a pre-existing defect.

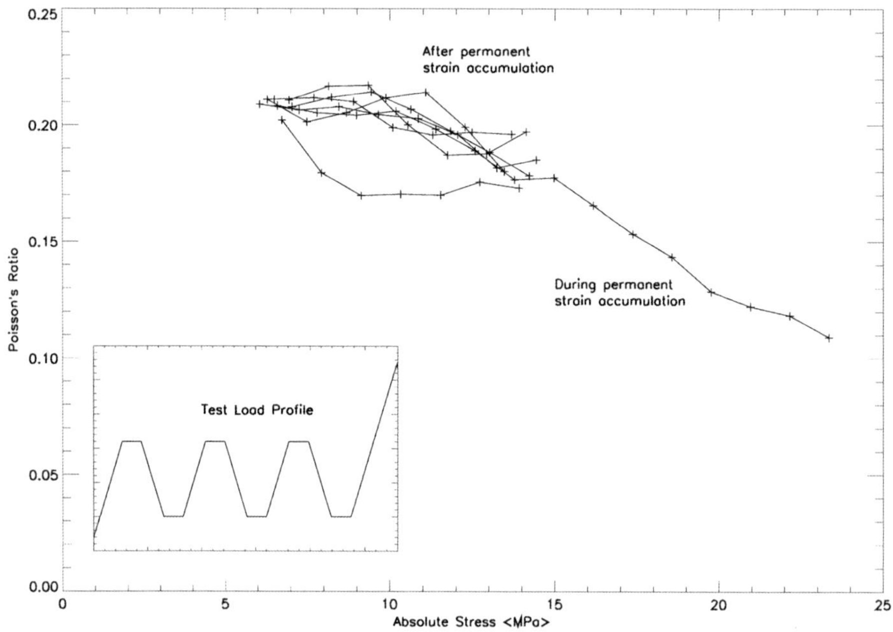

FIGURE 5: Poisson's ratio of virgin graphite beam sample subjected to cyclic loading.

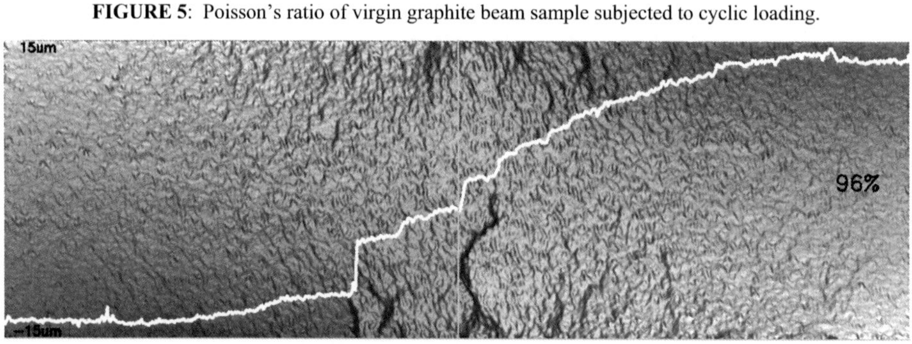

FIGURE 6: Response of virgin beam sample graphite to cyclic loading.

Irradiated Graphite

To date, ESPI has been used on 88 beam samples machined from AGR graphite. The values of Young's modulus obtained, when interpolated to 'zero' stress, agreed with the dynamic Young's modulus that was obtained using an ultrasonic time of flight technique. ESPI is able to show how the apparent 'static' Young's modulus varies with sample stress. Figure 7 shows that, unlike virgin graphite, the irradiated graphite modulus remains fairly constant up to about 50% of failure stress. There are no permanent sets in the stress-strain characteristics for irradiated Gilsocarbon graphite suggesting that the fast neutron irradiation has had a significant influence on the processes that cause these sets in unirradiated graphite.

Figure 8 shows how Poisson's ratio for Gilsocarbon graphite is related to sample stress for both virgin and irradiated graphite.

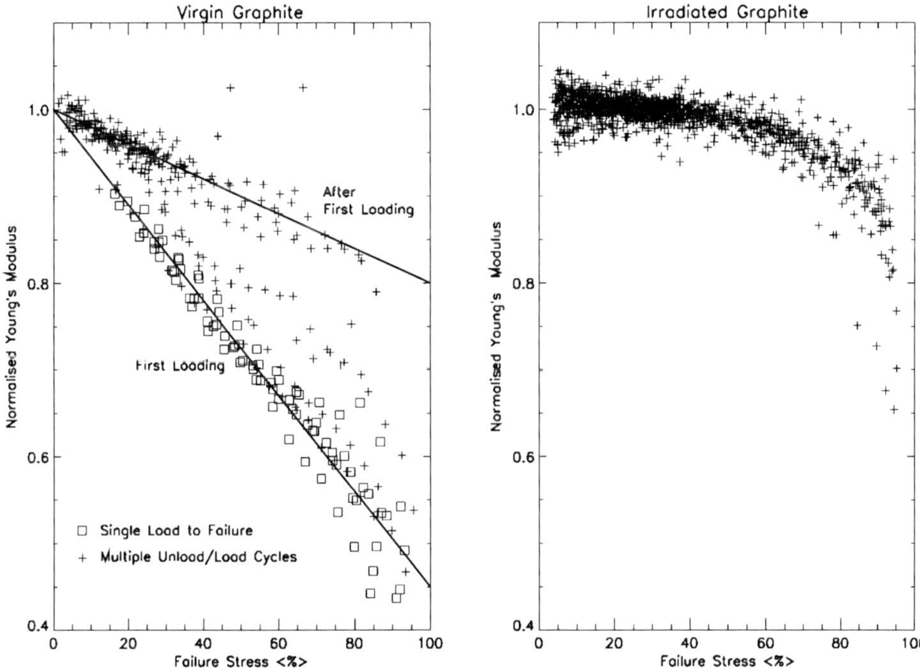

FIGURE 7: Young's modulus variation with stress in virgin and irradiated gilsocarbon graphite.

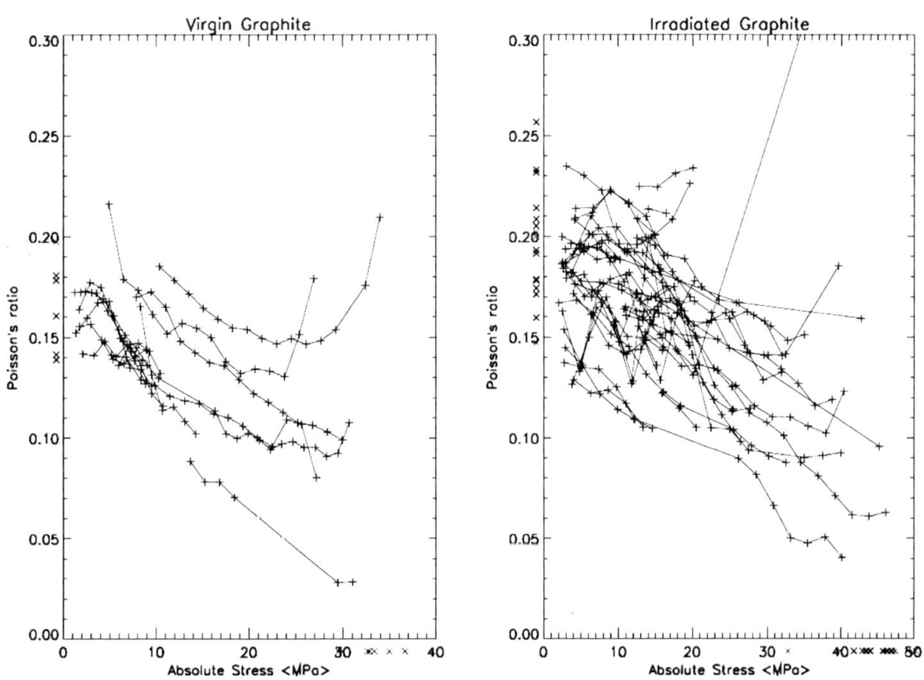

FIGURE 8: Poisson's ratio variation with stress in virgin and irradiated Gilsocarbon graphite.

CONCLUSIONS

- The use of ESPI as a non-contact measurement system has been successfully developed and is now being used for routine measurement of irradiated Gilsocarbon graphite and is significantly increasing the available data set for this material.
- Irradiated graphite shows minimal hysteresis during unload/load cycles.
- Irradiated graphite does not appear to experience any permanent damage at stresses below 50% of the failure stress.
- The measured value for Poisson's ratio is dependent on the absolute stress applied.
- Current data does not indicate that Poisson's ratio is changed significantly by irradiation and oxidation.

Acknowledgements

The author would like to thank EDF Energy for permission to present this paper and would like to make it clear that any views expressed are those of the author, not necessarily those of EDF Energy

References

Timoshenko, S. P. and Goodier, J. N. (1970). Theory of Elasticity, McGraw-Hill, p113-122.

Quantitative Microstructure Characterisation of Advanced Carbons Using Image Analysis

Theerapatt Manuwong[1], Gary D. Kipling[1], and Gareth B. Neighbour[2]

[1]Department of Engineering, University of Hull, Hull, UK. HU6 7RX
[2]Oxford Brookes University, Oxford, UK. OX33 1HX
Email: t.manuwong@2009.hull.ac.uk

Abstract
The complex structures and widespread industrial uses of advanced carbons, such as nuclear graphite, make it necessary to accurately characterise the material. Bulk properties of a material may be determined using mechanical testing, however, when dealing with a highly heterogeneous material with an irregular pore / flaw distribution, microstructural characterisation is advantageous. Image analysis provides a very powerful tool to determine and quantify microstructural features such as particle sizes, porosity, *etc.* This paper presents a novel characterisation methodology applied to advanced carbon materials using computational processes in conjunction with micrographs. In addition, the accuracy of the image data obtained is discussed in relation to validation.

Keywords
Characterisation, image analysis, porosity

INTRODUCTION

Nuclear graphite is used as a structural material as well as a moderator in nuclear reactors. During operation, nuclear graphite suffers from irradiation-induced damage to its microstructure, thus the microstructure deformation can lead to failure. Characterization of nuclear graphite material should be identified to understand the relationship between microstructure and bulk mechanical properties. Nuclear graphites have typical densities around 1.7-1.9 g/cm^3 which differs from the theoretical single crystal density of 2.265 g/cm^3, and so porosity is around 20% in nuclear graphites (Neighbour, 2010).

Porosity is the fraction of the total graphite volume that is taken up by the pore space. The graphite porosity plays an important role in the dynamics of fracture processes and favourably oriented pores may be the place of crack initiation under low stress level. On the other hand, crack propagation and growth may be influenced by the spatial relationship between large pores and pre-existing crack (Contescu, 2009). Although many types of pores, with various shape and size, are present in the graphite structure, it is generally agreed that large, slit-shaped pores are the most damaging to the graphite integrity (Burchell, 1996). Pore structure refers to the number and size of pores and to the interconnections between them which result from the packing of crystallites, particles, and disorganized carbon (Nightingale, 1962). A number of experimental methods have been used to study the pore structure. However, determination of the statistical distribution and direction of pores that is involved of the degradation leading to cracks and fracture in graphite is considered here. Quantitative data, attained from micrographs of graphite, are used to analysis pore size, distribution of pores and direction of pores. For example, the equivalent pore diameter (EPD) can be calculated by the equation

$$EPD = 2\sqrt{\frac{A}{\pi}} \qquad [1]$$

where A is the area of pore determined by image analysis software. The aim of this work is to use an image analysis methodology to observe the microstructure of advanced carbon, including Gilsocarbon graphite (IM1-24). Quantitative understanding of the characterisation of microstructure is important thus using quantitative data and knowledge gained from the microstructure especially pore size, distribution of pores and direction of pores are therefore important for probabilistic modelling of nuclear graphite studies in further work.

EXPERIMENTAL METHODOLOGY

The images of six difference pieces of IM1-24 graphite samples were taken using a 3.3 megapixel camera through the optical microscope. A total of 108 polarised light images from each sample were taken and arranged in a 9 x 12 rectangular grid arrangement. Each image was approximately 1.2 mm in width x 0.9 mm in length. Polarised light microscopy is an excellent tool to highlight specific constituents such as the filler, binder, *etc*.. Then image analysis software (Image-Pro Plus) was used coupled with a bespoke methodology applied to obtain the required data from the image. The methodology for extracting data from a raw optical micrograph is shown in Figure 1.

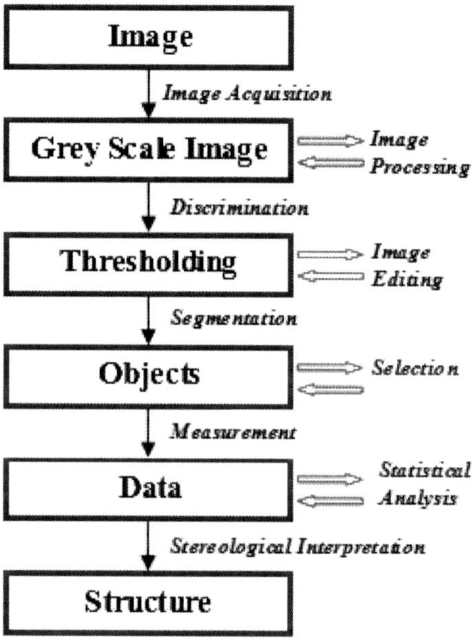

FIGURE 1: The overview of image analysis methodology (Russ, 1990).

First, an image of the material is generated using standard microscopy techniques (Russ, 1990). The tiling process generates a high resolution composite image which had an area of 10.8 mm x 10.8 mm, however; it was cropped down to 10 mm x 10 mm. When the composite image was converted into 8-bit grey scale, the threshold level allows the binary area of the composite image to be discriminated according to grey scale intensity. The segmentation range which was representing pores would have a lower limit of 0 thus the porosity areas which interested were relatively dark could be adjusted accordingly, see in Figure 2. However, the upper limit of this range should be validated due to it could easily vary between each image due to factors such as lighting, polarisation and exposure conditions of each image when it was taken.

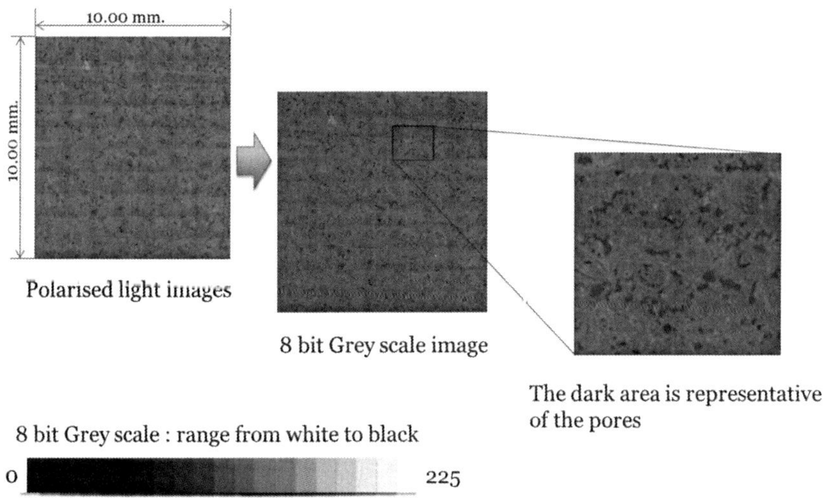

FIGURE 2: Shows the transition from composite polarised micrograph to the 8-bit grey scale converted image.

Segmentation is the key issue for image analysis. The appropriate threshold level should be used, to determine the suitable threshold level and so the specific methodology is required. In some case, the intensity between the boundary of the pore and surrounding are totally contrast due to a gradual brightness transition between the pore and the surrounding. The one method may assume the pore threshold level to be upper limit of the transition region by measuring the pore size (Wong, 2006).

Although the pore boundary may be atomically sharp, there is a gradual brightness transition between the pore and the surrounding material. Figure 3 shows this effect may be attributed to a difference in optical reflection of the epoxy resin at different 'depths'. The analogy of filling the pore with a fluid may be used, as mentioned by Wong (2006). The critical point is reached when liquid overflows into the surrounding areas, leading to a rapid increase in segmentation range. It is suggested that this 'overflow' point may provide a good estimate as to the pore threshold. The segmented objects are then used to characterise the porosity within the material. Data for individual pores such as area, aspect ratio, *etc.* is output allowing statistical characteristics to be determined. This technique can also be applied to other constituents such as filler particles.

RESULTS AND DISCUSSION

Image Pro Plus software was used to 'tile' the micrographs to generate a high resolution composite image with a resolution of approximately 220 megapixels. A grey scale threshold, based on the outlined methodology, was applied to the image and pore data was output. In the composite image, a total of 755,066 pores were measured and grouped into difference classes ranging in size from 1 μm^2 to 100,000 μm^2. Table 1 illustrates this with the vast majority of pores (~77.5%) being in Class 1 with an areal range between 1 and 10μm^2. Clearly, larger pore area classes have a lower number of objects.

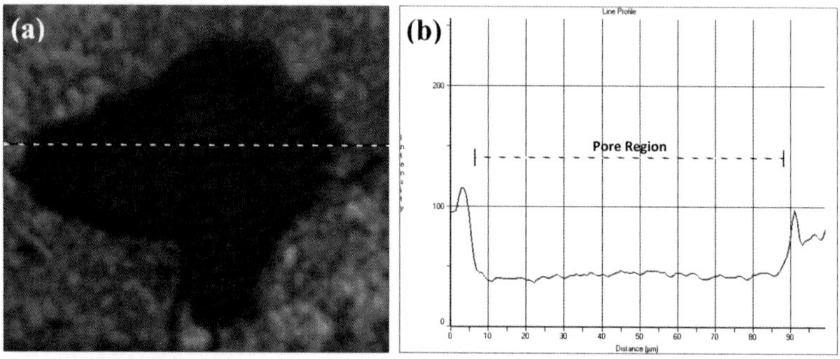

FIGURE 3: Pore depth: (a) grey scale line; and, (b) grey scale variation along horizontal line indicating porous region.

Class	Area Range (µm^2)	Objects	% Objects	Mean Area (µm^2)
1	1 to 10	584700	77.4369	3.7
2	10 to 100	157899	20.9119	24.9
3	100 to 400	8989	1.1905	179.4
4	400 to 1000	1691	0.2240	604.5
5	1000 to 5000	1233	0.1633	2280.8
6	5000 to 10000	285	0.0377	7128.2
7	10000 to 20000	179	0.0237	13930.7
8	20000 to 40000	73	0.0097	27304.8
9	40000 to 80000	16	0.0021	50123.0
10	80000 to 100000	1	0.0001	92295.9

TABLE 1: Data classes relating to pore sizes.

To better understand the statistical flaw distribution within the microstructure, Figure 4 shows a cumulative porosity curve against the equivalent pore diameter (EPD). The similarlity of the six distribution curves of the samples are shown. The same patterns of small pores distribution should be noted; all six samples show that they contain a vast amount of small pores with a high degree of similiarity. The larger pores are illustrated that their shapes are alike, even if the result of the Sample 4 is slightly deviated. It should be noted that numbers of medium to large pores contained in Sample 4 are more than the others. In additional, in all samples, the number of small pores is very large; however, they are less significant in terms of their contribution to overall porosity in the image. Conversely, large pores are few in number, but account for a substantial amount of the overall porosity. Their contribution to fracture is non-trivial (Kipling, 2011).

In Figure 5, pore colour coded of each class shows the distribution of all pores in graphite material. Figure 5(a) illustrates that all area of image are complicated and dense with varieties of size and different pore classes' illustarted. To ease this, it is perhaps useful, that the lower classes are removed to observe the larger pores in the image as see in Figure 5(b); the medium and large pore are clearly highlighted. The low density of large pores in upper class was identified when all of the lower classes were removed in Figure 5(c). Although, the large pores were few in number, but the location and shape account for a significant amount of the overall porosity. In contrast, the number of small pores was very large, however it shows less significance in terms of their contribution to overall porosity due to the size, location and shape. In some case the large group of small pores and the group location may also be significant.

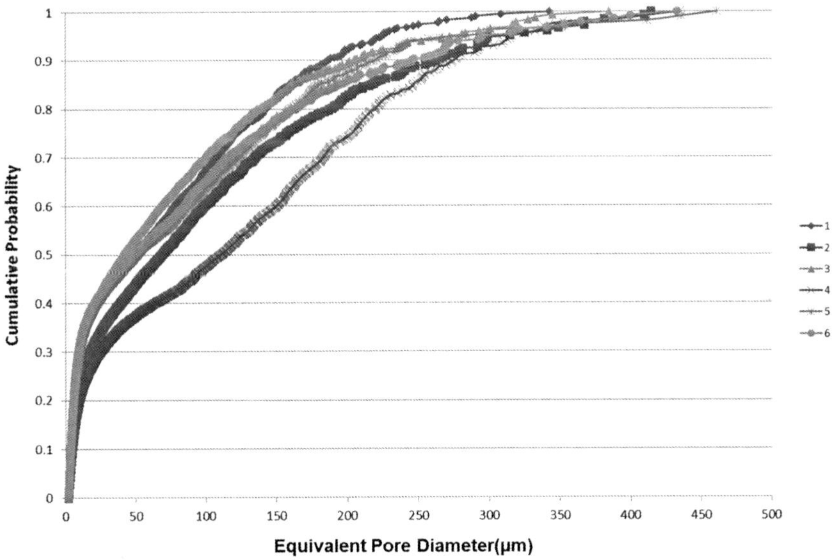

FIGURE 4: Pore size distribution for the six samples of IM1-24 graphite.

Furthermore the direction of all pores were observed, all pores were grouped in sets ranging by degree of angle. Figure 6 shows the histogram of the pores distribution by the degree of angle. There are 100 sets of pores, the first group start from 0 degrees to 1.8 degrees against the vertical line. Others are arranged in the same range and in order from 1.8 degree to 180 degrees, *i.e.* equi-spaced. The result should be considered; from a total of 228,384 objects, it is evident that the result that very fine pores tends to appear together and form a sort of pore clusters. Additional, the distribution shows the nearly symmetry of both sides of 90 degrees and also the majority of pores of the both sides are in the degree around between 25 degrees and 155 degrees (from 0 to 90 degrees and 90 to 180 degrees, respectively). Thus, the direction of the majority of pores and the pore cluster are significant which should be considered in term of crack propagation and failure.

(a)

Class	Area Range (μm²)	Objects	% Objects	Mean Area (μm)
	1 to 10	584700	77.4369	3.7
2	10 to 100	157899	20.9119	24.9
3	100 to 400	8989	1.1905	179.4
4	400 to 1000	1691	0.2240	604.5
5	1000 to 5000	1233	0.1633	2280.8
6	5000 to 10000	285	0.0377	7128.2
7	10000 to 20000	179	0.0237	13930.7
8	20000 to 40000	73	0.0097	27304.8
9	40000 to 80000	16	0.0021	50123.0
10	80000 to 100000	1	0.0001	92295.9

(b)

Class	Area Range (μm²)	Objects	% Objects	Mean Area (μm)
	1 to 10	0	0	0
2	10 to 100	0	0	0
3	100 to 400	8989	72.10235	179.42169
4	400 to 1000	1691	13.56381	604.4519
5	1000 to 5000	1233	9.89011	2280.7637
6	5000 to 10000	285	2.286035	7128.2393
7	10000 to 20000	179	1.435791	13930.727
8	20000 to 40000	73	0.585546	27304.816
9	40000 to 80000	16	0.128339	50123.027
10	80000 to 100000	1	0.008021	92295.875

(c)

Class	Area Range (μm²)	Objects	% Objects	Mean Area (μm)
	1 to 10	0	0	0
2	10 to 100	0	0	0
3	100 to 400	0	0	0
4	400 to 1000	0	0	0
5	1000 to 5000	0	0	0
6	5000 to 10000	285	51.44404	7128.2393
7	10000 to 20000	179	32.31047	13930.727
8	20000 to 40000	73	13.1769	27304.816
9	40000 to 80000	16	2.888087	50123.027
10	80000 to 100000	1	0.180505	92295.875

FIGURE 5: The pore colour coded images shows in (a) all of the classes (class 1-10); (b) the class 3-10; and (c) the class 5-10.

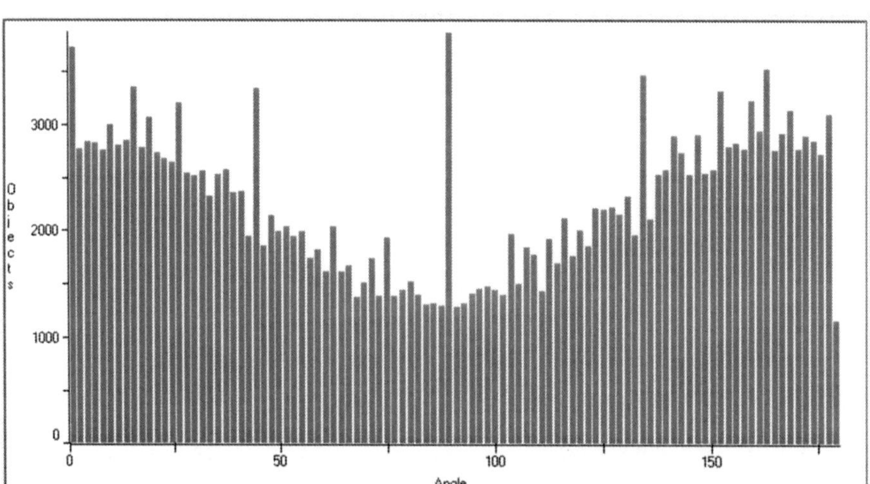

FIGURE 6: Distribution histogram between pores and their direction.

CONCLUSION AND FUTURE WORK

The use of image analysis, such as the methodology outlined here, allows for a greater understanding as to composition of the material. Statistical flaw distributions may be output to quantifiably characterise the material. This in turn, may aid in the explanation of material properties, fracture and failure analysis. Further work will help to characterise the material in greater depth to predict performance, *e.g.* irradiation creep, through a structured programme incorporating particle and pore shapes and sizes sampled at higher resolutions.

References

Burchell T. D. (1996). A microstructurally based fracture model for polygranular graphites. *Carbon,* 34, 297-316.

Contescu, C. I. & Burchell, T. D. (2009). Charaterization of porosity development in oxidized graphite using automated image analysis techniques. ORNL/TM-2009/192, Oak Ridge National Lab., Oak Ridge.

Kipling, G. & Neighbour, G. (2011). Novel FEA methods for predicting fracture characteristics in nuclear graphites. Carbon 2011 Proceedings, International Conference on Carbon, Shanghai, China.

Neighbour, G. B. (2010). Chapter 23 - Plant life management (PLiM) practices for gas-cooled, graphite-moderated nuclear reactors: UK experience. In:- Understanding and mitigating ageing in nuclear power plants: Materials and operational aspects of plant life management (PLiM) (Editor: Philip G. Tipping), Woodhead Publishing, Cambridge, UK, 838-875. ISBN 978-1-84569-511-8.

Nightingale, R. E. (1962). Nuclear graphite. London: Academic Press Inc. Ltd.

Russ, J. C. (1990). Computer-Assisted Microscopy: The Measurement and Analysis of Images. New York, Plenum Press.

Wong, H. S., Head, M. K., Buenfeld, N. R. (2006). Pore segmentation of cement-based materials from backscattered electron images. *Cement and Concrete Research*, 36, 1083-1090.

Revised Method For Graphite Weight Loss Prediction

Jonathan E. Coote and Neil S. Headings
Frazer-Nash Consultancy, 1 Lower Lamb Street, Bristol, UK. BS1 5UD
Email: j.coote@fnc.co.uk

Abstract
In this paper, the authors describe a new approach for predicting Gilsocarbon graphite weight loss which enables variability and uncertainty of predictions to be calculated from a physical model. The FEAT-GRAPHITE code is used to model weight loss, density and bore diameter through time as a function of various input parameters. The input parameters are defined as probability distributions, and the distributions are optimised in order to fit predictions to the spread of observed values. Uncertainty is quantified via bootstrapping of the measurement data. The method is presented and illustrated with sample results.

Keywords
Shrinkage, variability, uncertainty

INTRODUCTION

A key requirement of many safety cases for Advanced Gas-cooled Reactors (AGR) is the accurate forward prediction of weight loss within the graphite bricks which constitute the structure of the reactor core. The FEAT-GRAPHITE finite element code (Hickmott *et al.*, 2008) has been developed which uses physical models to predict the properties of graphite bricks within the core. Broader description of graphite ageing may be found in Neighbour *et al.* (2007). Forward predictions may address specific samples of material (*e.g.* trepanned samples at a given outage), or may instead be forecasts of derived quantities (*e.g.* average overall core weight loss, key failure probability, or probability that a given number of bricks throughout the core will experience weight loss beyond a threshold value over a certain proportion of their volume).

For clarity, trepanned graphite cores shall be referred to as 'samples', each of which are cut into 'slices'; the measurement database consists of values of density and weight loss per slice, and values of bore diameter per sample.

FEAT-GRAPHITE accepts two input parameters which are particularly suitable for model tuning: Attack Rate Multiplier (ARM) modifies the rate of radiolytic oxidation, and A_{scale} adjusts the rate at which dimensional change takes place. Additionally, a channel power factor, F, is used to adjust the burnup as appropriate to the given channel. Given scalar values for these parameters, FEAT-GRAPHITE may deterministically calculate density, weight loss and bore diameter over the brick mesh through time. A subsequent inspection of the mesh results yields values of density and weight loss for any given slice, and bore diameter for any given sample. Distributions of virgin bore diameter and virgin density may be specified, capturing the known variability of these input parameters; these are used to adjust the predicted values of bore diameter and density in an additive sense when optimising ARM and A_{scale}.

The FEAT-GRAPHITE physical input parameters were originally tuned against measurements made in various materials test reactors (MTRs). However, discrepancies were found between the resulting predictions and measurements at operating AGRs. Variability in the graphite production process and operating conditions has resulted in material properties differing from reactor to reactor, from brick to brick, and within individual bricks. MTR measurements are unable to reproduce the true variability observed within the operating stations. It is therefore necessary to modify the FEAT-GRAPHITE inputs using empirically-determined reactor-specific calibration factors in order to obtain more accurate predictions.

Previous work has focussed on using a post-hoc statistical correction to the results of FEAT-GRAPHITE to account for the inter-brick variations in the oxidation rate. This technique has been termed the 'back-end' methodology. However, the technique has a number of limitations and does not separate variability from prediction uncertainty. This paper describes the approach taken by an alternative, 'front-end', methodology.

METHODOLOGY

Overview

To improve upon the back-end methodology a new technique termed the 'front-end' methodology has been formulated which can: (i) predict measurable endpoints such as bore diameter and density; (ii) predict values for trepanned samples without reference to measurements made at the time of inspection; and, (iii) propagate the variability and uncertainty in predicted quantities through the prediction process.

The front-end methodology describes the natural variability in the graphite material by describing ARM and A_{scale} using probability density functions. Their associated distributions are assumed to take an analytical form defined by a set of characteristic parameters, such as the mean and standard deviation for a normal distribution. The distribution parameters are optimised to measured data from a series of trepanning campaigns. The uncertainty in the fitted distribution parameter values is assessed by a bootstrap approach. The distribution of the retuned parameters is noted and describes the uncertainty. Having established a method for predicting endpoint values for individual samples, it is possible to formulate probabilistic forecasts of derived quantities.

FEAT-GRAPHITE Emulation

A large number of model evaluations are necessary during the optimisation stage; however, as each FEAT-GRAPHITE evaluation consists of a full through-life finite element analysis, the computational cost of performing this analysis becomes prohibitive. Therefore, a series of response functions has been fitted to a database of stock results sampling the parameter space of ARM & A_{scale} values, thereby allowing rapid evaluation of predictions.

The response surfaces emulation for a FEAT-GRAPHITE prediction of bore diameter, ϕ_i, for a sample i is given by:

$$\phi_i = \tilde{f}_{\phi_i}(\kappa_i, C_S, A_S, F) = \sum_{l,m,n} \kappa_{i_{lmn}} C_s^l A_s^m F^n \qquad [1]$$

where $\tilde{f}_{\phi_i}$ is the bore diameter polynomial response function per trepanned sample, $\kappa_{i_{lmn}}$ are a set of fitted polynomial coefficients, unique to each sample, F is the channel power factor, and C_s and A_s are A_{scale} and ARM respectively. The polynomial coefficients describing the response function are defined for specific values of core burnup. Response functions at other burnup values are defined with respect to interpolated FEAT-GRAPHITE predictions.

Likewise, the response function for the density of the jth slice within sample i is given by:

$$\rho_{ij} = \widetilde{f}_{\rho_{ij}}\left(\eta_{ij}, C_S, A_S, F\right) = \sum_{l,m,n} \eta_{ij_{lmn}} C_s^l A_s^m F^n \qquad [2]$$

where $\widetilde{f}_{\rho_{ij}}$ is the density polynomial response function and $\eta_{ij_{lmn}}$ are the set of polynomial coefficients unique to each slice. For a polynomial of order n, the multi-dimensional polynomials will initially have $(n+1)^3$ coefficients, although the number is then reduced during a process of stepwise regression when insignificant terms are eliminated. Typically, the response surfaces are observed to be approximately linear and vary smoothly; the error between FEAT-GRAPHITE values and response surface values is insignificant in comparison with the likely magnitude of measurement error.

Optimisation Process for ARM and A_{scale} Distributions

The parameters that describe the distributions of ARM and A_{scale} are optimised by minimising the negative log-likelihood that the parameters would result in the set of observed measurements. This requires the evaluation of probability density functions (pdfs) for the observed quantities (density, bore diameter) for a given distribution of ARM and A_{scale}. These pdfs are approximated by Monte Carlo sampling from the ARM and A_{scale} distributions (assuming independence) and using the response functions for each slice to generate a distribution of predictions. A likelihood value per measurement is found from the distribution of corresponding predictions by using a kernel density estimate of the probability density at the measured value. The sum is taken across all negative log-likelihoods, and minimised to find optimal parameters for ARM and A_{scale}. The choice of random numbers is consistently seeded within the Monte Carlo samplings in order to ensure that the computed likelihood values are repeatable, assisting optimisation. A particle swarm optimisation precursor stage is carried out in advance of a simplex method in order to reduce the possibility of solution capture within a local optimum.

Optimisation Process Modification for Capturing Brick-to-Brick Variability

The optimisation as described above captures variability from slice to slice as well as from sample to sample and brick to brick. It would be non-conservative to use weight loss predictions based on this approach within the keyway cracking structural integrity analysis as a narrower distribution is more onerous and results in predicting more rapid progression of brick cracking throughout the core. The optimisation has therefore been modified to predict sample-average values rather than slice-specific values. The sample-average probability distribution is constructed by Monte Carlo sampling: per iteration, values are sampled for every slice in each trepanned sample and a weighted mean per sample is calculated. The weighting function may be arbitrarily chosen; weighting by slice thickness may be desirable. The probability distribution of the weighted mean may then be defined based on the sampled values, *e.g.* by using a kernel density estimator. An equivalent process may be used to calculate the probability distribution for brick-average values. In cases where the database of measurements does not contain many multiply-trepanned bricks, the distribution of brick-average values will be close to the distribution of sample-average values.

Choice of ARM and A_{scale} Distributions

While arbitrary distributions may be defined for each of ARM and A_{scale}, it appears likely that common 2- or 3- parameter distributions shall suffice. In order to assess the relative merit, the Akaike information criterion may be used to rank alternative choices in accordance with the maximised value of the product of the likelihoods balanced against a penalty that increases with the number of parameters.

Quantification of Uncertainty

The method thus far allows distributions to be calculated for predicted quantities (*e.g.* density of a given slice at a particular point in time), each with an associated variability. The variability accounts for fluctuations in virgin bore diameter, virgin density, oxidation rate and shrinkage rate. However, there also exist a number of sources of uncertainty. Systematic sources include: uncertainty in FEAT-GRAPHITE inputs (*e.g.* dosimetry, temperature) which are presently modelled deterministically, model uncertainty (*e.g.* choice of assumed distribution family for ARM), and random measurement error. Quantification of these systematic sources is not straightforward. Random measurement error is not separately modelled, instead contributing to fitted variability. Model uncertainty is not straightforward to quantify other than by repeating the analysis with alternative assumptions. A further source of uncertainty arises due to the limitations on the amount of measurement data available. This final source is considered by a further stage of the analysis, now described.

To evaluate the uncertainty in predictions resulting from limited numbers of measurements, the ARM & A_{scale} tuning process is bootstrapped across a large number of resampled measurement datasets. Each resampled set of measurements contains the same number of elements as the original, with each new member chosen from the set of original members without bias or replacement. The repeated tunings produce a distribution for each input parameter (*e.g.* distribution of tuned ARM mean, distribution of tuned A_{scale} standard deviation). The standard deviation of such distributions is an indication of the uncertainty associated with each parameter. Distributions of predictions for measured quantities may then be calculated through Monte Carlo sampling across the set of ARM and A_{scale} distributions rather than just the tuned distributions of each (assuming independence between the distribution parameters). Such distributions account for both uncertainty and variability, and may be compared against those which consider variability only. It is expected that the uncertainty around the ARM and A_{scale} parameters will reduce as the measurement dataset is expanded.

It will be noted that each tuning stage involves a Monte Carlo process to be repeated for each measurement for each evaluation of the overall likelihood; the implementation is therefore significantly parallelised for expediency.

Examination of Correlations

The methods described thus far assume that there is no significant correlation between the A_{scale} and ARM distributions, or more specifically, that the variability of each parameter is uncorrelated. Also of interest is whether the uncertainty of the distribution parameters is correlated. These correlations can be evaluated in two further stages of investigation.

To check for the possible existence of correlated variability, a new fitting is carried out for A_{scale} and ARM, this time fitting a scalar value of each to every slice (or sample, if desired). This results in a set of A_{scale} and ARM value pairs, and the correlation between the two quantities may then be assessed. The fitting is carried out so as to minimise the sum of squared normalised residuals of the predicted quantities (density, weight loss and bore diameter). For the normalisation, the standard deviation of each predicted quantity is used. The standard deviations are in turn evaluated by Monte Carlo sampling from the tuned A_{scale} and ARM distributions and using the response functions to assess the distribution of predicted quantities. The distribution of A_{scale} and ARM obtained in this way may also be compared to the assumed forms used during the main tuning as a cross-check. To check for the possible existence of correlated uncertainty, the series of A_{scale} and ARM means obtained from the bootstrap analysis may be assessed visually and numerically.

SAMPLE RESULTS

Some sample results are provided to illustrate aspects of the process. Example AGR bore diameter, density and weight loss measurements are shown in Figure 1. FEAT-GRAPHITE fitted response functions are shown for a single slice and burnup in Figure 2. Figure 3 shows an example of tuned FEAT-GRAPHITE input parameter distributions, while Figure 4 shows the variation of the tuned parameters across bootstraps.

Figure 5 shows how predictions may be compared against measurements, while Figure 6 demonstrates how the stability of the tuned parameters may be reviewed in light of an expanding measurement database. Figure 7 shows how a forecast of peak brick average weight loss through time may be presented, showing the $5^{th}/50^{th}/95^{th}$ variability percentiles and their associated uncertainties. Presentation of the results in this way promotes interpretation of the analysis findings and offers improvement over incumbent methods.

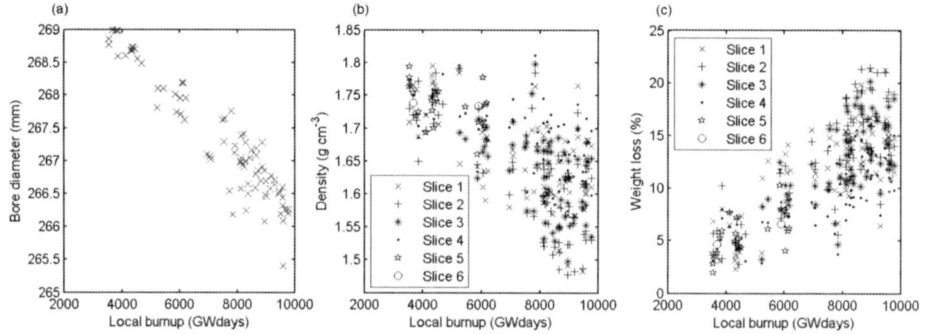

FIGURE 1: Example AGR measurement data.

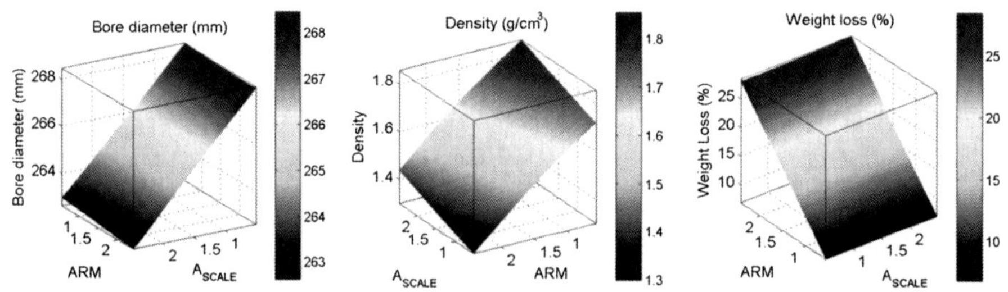

FIGURE 2: Example response functions for a given slice/burnup.

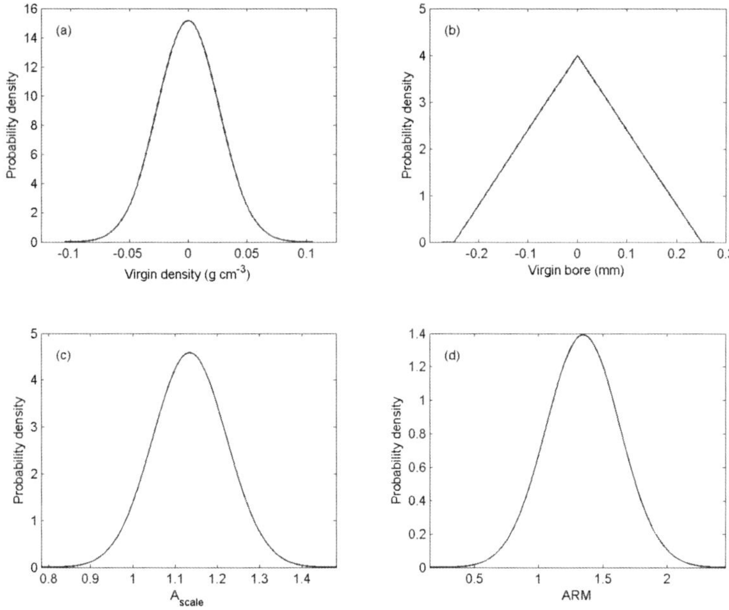

FIGURE 3: Example input parameter distributions following optimisation.

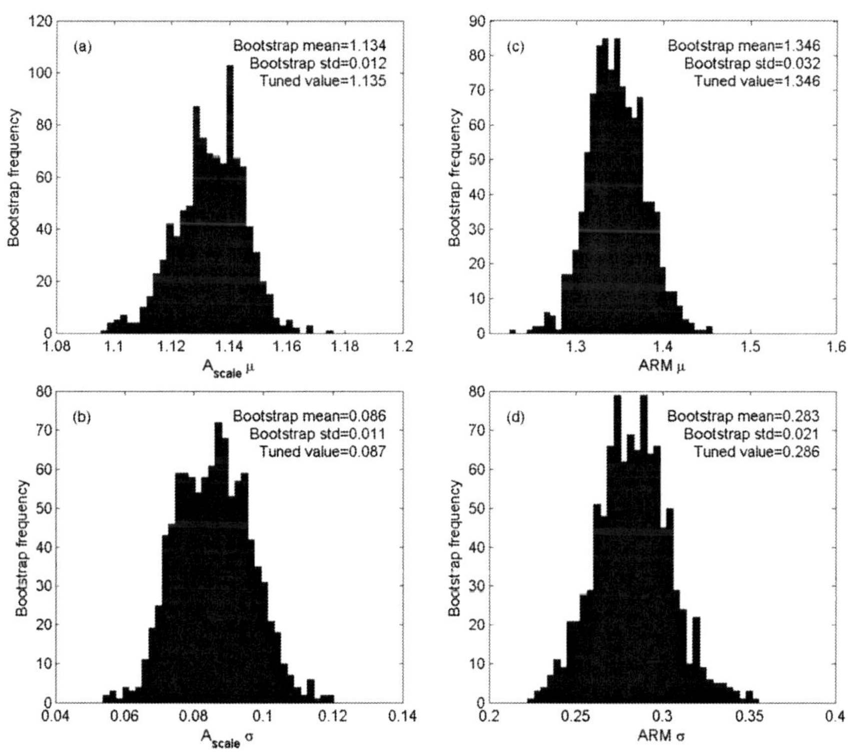

FIGURE 4: Example bootstrap results for A_{scale} and ARM distribution parameters.

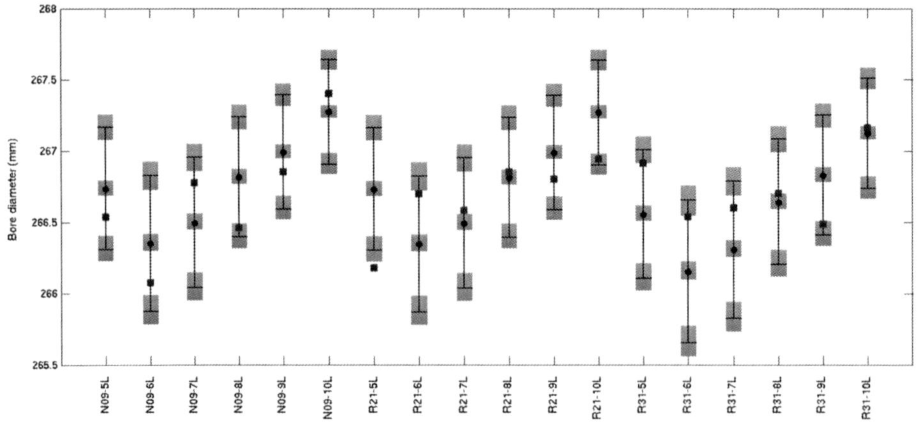

FIGURE 5: Example comparison of predictions against measurements. Black bars indicate 5% and 95% bound, circles indicate best-estimate predictions, black squares indicate measured values, and grey zones indicate 5%-95% uncertainty on the bounds.

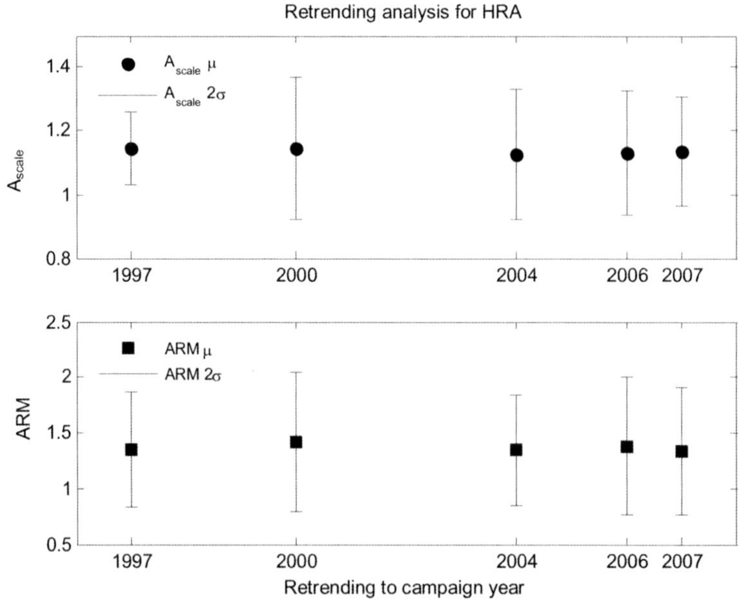

FIGURE 6: Example retrending analysis showing effect on tuned input distributions of introducing data from successive trepanning campaigns. Bounds indicate variability; uncertainty not illustrated.

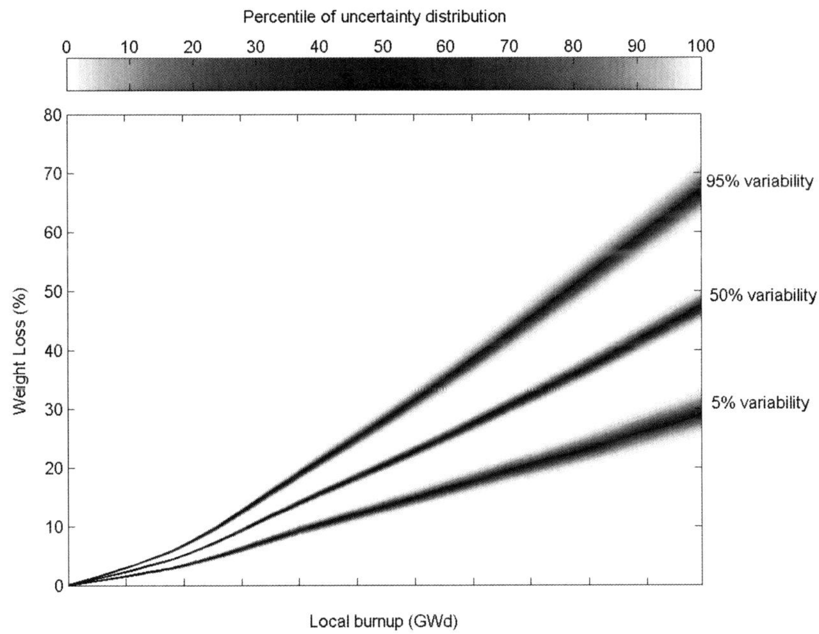

FIGURE 7: Example forecast of volume-averaged weight loss over peak brick through life, showing 5th, 50th & 95th variability percentiles, each with shaded uncertainty distribution.

CONCLUSIONS

A method based on physical modelling has been implemented by which predictions of measured density, weight loss, and bore diameter may be made. Abstraction of the physical model via response functions has rendered feasible the use of stochastic methods for assessing proposed distributions of FEAT-GRAPHITE input parameters. The method allows material variability and random uncertainty to be quantified separately and propagated through the process of forecasting derived quantities. When combined with suitable visualisation techniques, the method promotes improved understanding of the evolution of core properties, their variability, and how much confidence may be placed in the results.

Acknowledgements
The authors would like to thank EDF Energy for financial support and permission to present this paper. The authors acknowledge helpful advice received from Peter Robinson and Philip Maul of Quintessa, Mark Bradford and Peter Matthews of EDF Energy, and Samantha Dugelay of Frazer-Nash Consultancy. Any views expressed in this paper are those of the authors and may not represent those of EDF Energy or Frazer-Nash Consultancy.

References

Hickmott, S., Smith, R. M., Rabbitt, M. J. and Fairbairn, S. A. (2009). FEAT-GRAPHITE: Graphite Weight Loss and Stress Prediction, Theory Manual and User Guide. British Energy Internal Document.

Neighbour, G. B. (Editor) (2007). Management of Ageing in Graphite Reactor Cores. RSC Publishing (Royal Society of Chemistry, Cambridge). ISBN-10: 0854043454; ISBN-13: 978-0854043453.

On Large Scale Implicit Multibody Contact Dynamics Modeling with SOLFEC

Tomasz Koziara and Nenad Bićanić
School of Engineering, University of Glasgow, Rankine Building, Glasgow, G12 8LT.
Email: t.koziara@gmail.com, nenad.bicanic@glasgow.ac.uk

Abstract
SOLFEC is a multibody contact/impact simulation code. Unlike most of the currently used explicit multibody codes, it does not use repulsive springs to prevent interpenetration. Instead, it exploits an implicit formulation of contact conditions, where physically plausible impact resolution follows from momentum balance and deformations of contacting bodies. As a result, stiff springs are eliminated and larger stable time steps are allowed. In this contribution, we outline the basic principles behind the computational approach employed in SOLFEC. We give examples relevant to dynamic analysis of AGR graphite cores, demonstrating that our approach can effectively deal with multibody structures composed of thousands of loosely keyed together bodies with tight geometrical features and small clearances. Multiple impacts, disengagement of interacting features, and fragmentation of bodies are all handled in our implementation. These aspects, as well as the parallel efficiency will also be exemplified. SOLFEC can be downloaded from: http://code.google.com/p/solfec/

Keywords
Multibody dynamics, friction, high performance computing

INTRODUCTION

In the field of dynamic multibody contact and impact analysis the current state of the art dates back to Cundall and Strack (1979) and the Discrete Element Method (DEM). This is essentially an approach based on repulsive springs, preventing excessive interpenetration between overlapping bodies. This approach proved to be useful in many practical applications (*e.g.* granular flow) and it is now one of the standard tools in engineering applications.

In a typical DEM implementation, an analytical relation between the interpenetration depth and the contact force (*e.g.* the Hertz contact model) is augmented with a damper element and further combined with the rigid body model and explicit integration in time. For simple shapes such as spherical or ellipsoidal pebbles, this usually well reflects the actual physics. In case of complex (*e.g.* non-convex, polyhedral) shapes the spring-damper contact model becomes a purely numerical device. In such cases careful tuning is required in order to fit with experiments, which usually results in very stiff interface laws. As a consequence, the stable time step is typically brought down below the level of microseconds, which may be inefficient for long-term simulations (*e.g.* with many thousands deformable bodies). Additionally, such stiff interface laws introduce high frequency modes, which are most often of no practical interest.

It is then relevant to think of a multibody formulation that can effectively skip over the high frequency content and directly deal with the slower dynamics, which is typically of interest in structural applications. In this context, our point of departure is the Contact Dynamics (CD) method developed by Moreau (1999) and Jean (1999). This approach does not rely on a closed form interpenetration-force relation, but rather it employs a complementarity relation between the relative velocity and the contact force at an existing contact point: either the velocity is such that the bodies separate and the force is zero, or the force is such that interpenetration is prevented and in consequence the relative velocity is zero. This velocity-force complementarity relation is added as an algebraic constraint to the implicitly integrated momentum balance. Hence, a nonlinear implicit problem is solved at every time step in order to find the contact forces and the velocities of bodies. This is time consuming, but since the time step can be large, in many cases the overall runtime may be significantly shorter than in the case of an equivalent DEM simulation.

SOLFEC implements an instance of the Contact Dynamics approach. It allows to model dynamic multibody systems with joints, contacts/impacts and friction, using a variety of kinematic models (*e.g.* rigid, pseudo-rigid, finite element). Currently the code is optimized to effectively simulate problems with thousands of interacting bodies, each having up to several thousands of degrees of freedom, and with arbitrary polyhedral, spherical or ellipsoidal shapes. In order to achieve high efficiency, SOLFEC exploits the Message Passing Interface (MPI) and runs on distributed memory supercomputers.

In the remaining part of the paper we give a brief summary of the Contact Dynamics formalism, followed be several illustrative examples generally relevant in the context of the dynamic analysis of AGR graphite cores. Finally, we conclude and comment on the ongoing and future work.

SUMMARY OF THE CONTACT DYNAMICS APPROACH

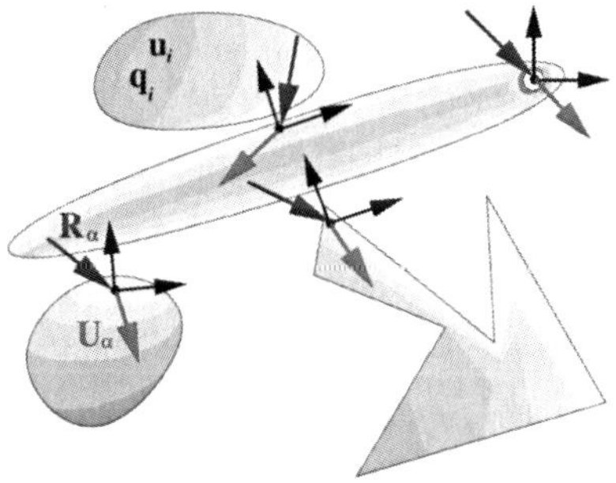

FIGURE 1: A multibody system.

At time t a set of bodies has configuration $\mathbf{q}^t$ and velocity $\mathbf{u}^t$ (cf. Figure 1). Consecutive values $\mathbf{q}^{t+h}$, $\mathbf{u}^{t+h}$ for $h>0$ are obtained by numerical integration of the momentum balance and the kinematic relation $\dot{\mathbf{q}}=\mathbf{u}$. A half-step configuration is obtained as:

$$\mathbf{q}^{t+h/2}=\mathbf{q}^t+\tfrac{h}{2}\mathbf{u}^t. \qquad [1]$$

The velocity update reads:

$$\mathbf{u}^{t+h}=\mathbf{b}^{t+h}+\mathbf{A}^{-1}\mathbf{H}^T\mathbf{R}, \qquad [2]$$

where $\mathbf{b}^{t+h}=\mathbf{u}^t+h\mathbf{A}^{-1}\mathbf{f}$ is a free velocity, not accounting for an influence of constraints. When $\mathbf{A}=\mathbf{M}$, where $\mathbf{M}$ is an inertia matrix, Equation [2] corresponds to an explicit scheme suggested in Moreau (1999). For $\mathbf{A}=\mathbf{M}+\frac{h^2}{4}\mathbf{K}$, where $\mathbf{K}=-\{\partial\mathbf{f}/\partial\mathbf{q}\}(\mathbf{q}^{t+h/2})$ is a stiffness matrix, Equation [2] corresponds to the linearly implicit scheme by Zhang and Skeel (1997). In both cases, the out of balance force $\mathbf{f}=\mathbf{f}(\mathbf{q}^{t+h/2},t+\frac{h}{2})$. For smooth problems, both integrators have good conservation properties and are of second order. Naturally, the linearly implicit scheme allows for larger time steps. While $\mathbf{b}^{t+h}=\mathbf{u}^{t}+h\mathbf{A}^{-1}\mathbf{f}$ represents a regular velocity update for an unconstrained system, Equation [2] accounts for the constraints by adding a contribution of their reactions $\mathbf{R}$. In Equation [2], $\mathbf{A}=\mathbf{A}(\mathbf{q}^{t+h/2})$ and $\mathbf{H}=\mathbf{H}(\mathbf{q}^{t+h/2})$, while $\mathbf{R}=\int_t^{t+h}d\mathbf{R}$ is considered to be an impulse. $\mathbf{H}$ is a linear operator mapping the body space velocity $\mathbf{u}^{t+h}$ into a constraint space velocity $\mathbf{U}$

$$\mathbf{U}=\mathbf{H}\mathbf{u}^{t+h} \quad [3]$$

For example, in case of finite elements, $\mathbf{H}$ is a global to local coordinates transformation of the shape functions values nonzero at constraint points. Once $\mathbf{R}$ is known, after obtaining $\mathbf{u}^{t+h}$ from Equation [2], the final configuration update reads:

$$\mathbf{q}^{t+h}=\mathbf{q}^{t+h/2}+\tfrac{h}{2}\mathbf{u}^{t+h} \quad [4]$$

In order to compute $\mathbf{R}$, Equations [2] and [3] are solved together with a set of nonlinear implicit relations of a general form:

$$\mathbf{C}_\alpha(\mathbf{U}_\alpha,\mathbf{R}_\alpha)=\mathbf{0}, \quad [5]$$

where $\mathbf{U}_\alpha$ and $\mathbf{R}_\alpha$ are the 3-component relative contact velocities and contact reactions at individual contact points. In the most typical case, Equation [5] combines the velocity level Signorini condition with the Coulomb friction law, *cf.* Figure 2. It should be noted that the above summary exemplifies only a particular instance of CD, as implemented in SOLFEC (*cf.* Koziara and Bićanić (2011) and references therein).

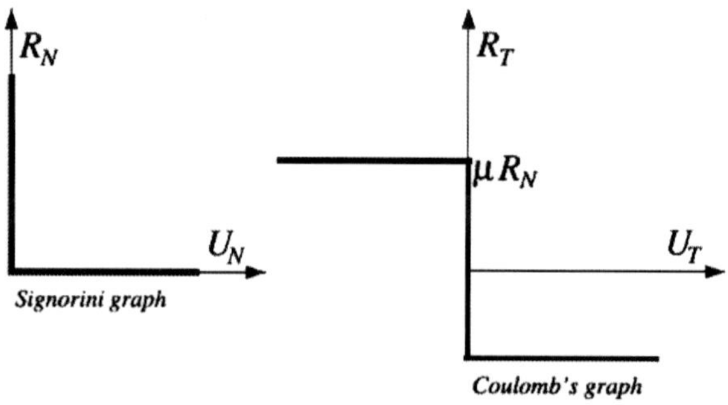

FIGURE 2: The Signorini-Coulomb frictional contact law, where R_N, U_N, R_T, U_T are the normal and the tangential components of the contact reaction and relative velocity, and μ is the coefficient of friction.

EXAMPLES

The examples given in this section illustrate the potential capability of SOLFEC to model dynamics of AGR graphite cores. The employed geometry is similar to that of AGR cores, but it does not correspond to actual core models.

Modeling Shapes and Deformations

The polyhedral geometry in SOLFEC can be created in two ways: it can either be a typical FEM mesh (comprising hexahedrons, tetrahedrons, pyramids and wedges), or it can be a juxtaposition of arbitrary convex polyhedrons. In Figure 3, a shape similar to a fuel brick is juxtaposed out of several dozens of simple convex objects. For this sort of shape, three different deformation models are available:

1. Rigid body model.
2. Pseudo-rigid body model, where global linear deformations are only allowed.
3. Background FEM mesh model, where any background FEM mesh can parameterize deformations of the shape.

FIGURE 3: Shape similar to a fuel brick, modeled as juxtaposition of simple convex objects. In the middle the shape is deforming according to the linear pseudo-rigid motion. On the right the shape is deforming in accordance with a background hexahedral mesh.

Fragmentation

Modeling fragmentation of bodies during a dynamic simulation may be a desirable feature. SOLFEC includes a simple fragmentation model, that allows predefining potential fragmentation planes, together with a fragmentation criterion (*e.g.* maximal tensile stress). Since accurate modeling of dynamic cracking in large multibody structures would be extremely challenging, our simple fragmentation model allows to incorporate a preexisting expert knowledge and run analyses based on various "what if" scenarios. Figure 4 illustrates fragmentation after a drop of a simplified single fuel brick model.

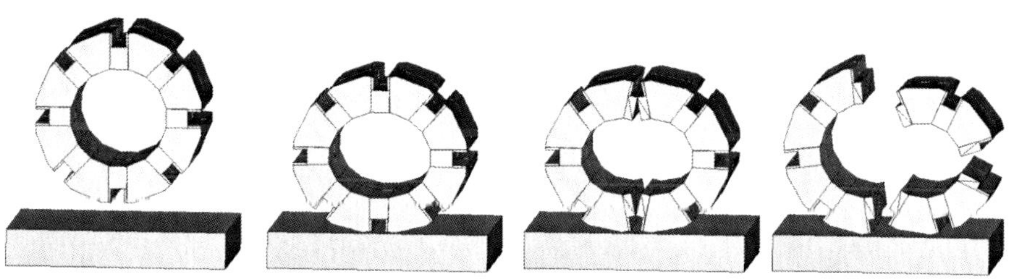

FIGURE 4: Drop and fragmentation of a simplified fuel brick model with two predefined crack planes (a double and a half crack produce three fragments).

Box-Kite Push Until Lockup

One of our concerns at early stages of SOLFEC development was whether a velocity driven CD approach (as opposed to the interpenetration driven approaches) could yield sufficient accuracy of contact resolution for multibody simulations with tight clearances between geometrical features. A box-kite test example was developed, mimicking a quasi-static experiment where a 3x3 array of flat acrylic bricks was pushed until lockup. One of the models is illustrated in Figure 5. A good general agreement was found between the experiment, SOLFEC and a previous FEM model. The general feature of the implicit approach implemented in SOLFEC is that although interpenetrations may occur, they do not grow once contact has been established. Hence the amount of the interpenetration is proportional to the amount of possible motion during one time step. Typically this can be kept well below the level of the needed accuracy (*e.g.* 0.1 mm in the box-kite test).

FIGURE 5: The box-kite test, where the middle doubly cracked bricks are sheared until lockup.

Simplified Quarter Core

Figure 6 illustrates the simplified quarter core model, developed in order to test the computational efficiency of the parallel version of SOLFEC. The model is based on a repetition of a pattern of loosely keyed together bricks and it is meant to approximate the computational complexity of a more elaborate model. The timings in Figure 7 correspond to a model based on 4000 pseudo-rigid and 1333 rigid bodies (55998 degrees of freedom), subject to gravity and resting on a rigid foundation. The horizontal acceleration of the foundation equals to a seismic signal. 1 second runs were computed with the time step 1E-4 in order to test the parallel scaling on 8 to 64 processors. Additionally, the 64 processor run was repeated with time steps 2E-5, 5E-4 and 1E-3 in order to exemplify the influence of the step size on total runtimes. The tests were run on an Intel Xeon 5472 3.00GHz based cluster with Infiniband network. In Figure 7, we can see that a super-linear scaling can be observed and that for larger times steps the total runtimes grow and are dominated by the implicit solution of the constraints equations (*i.e.* it is harder to resolve contacts and other constraints if a larger amount of relative motion is allowed during a time step).

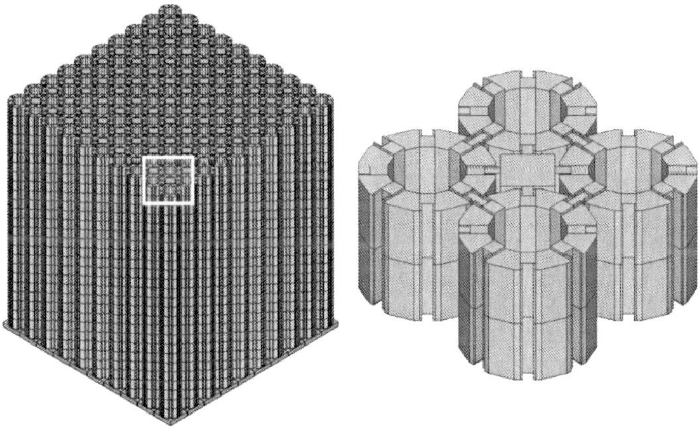

FIGURE 6: The simplified graphite core example. The model comprises twelve layers of a repeated pattern, detailed on the right. The layers are connected through convex-concave features of the top and bottom surfaces of the hollow bricks.

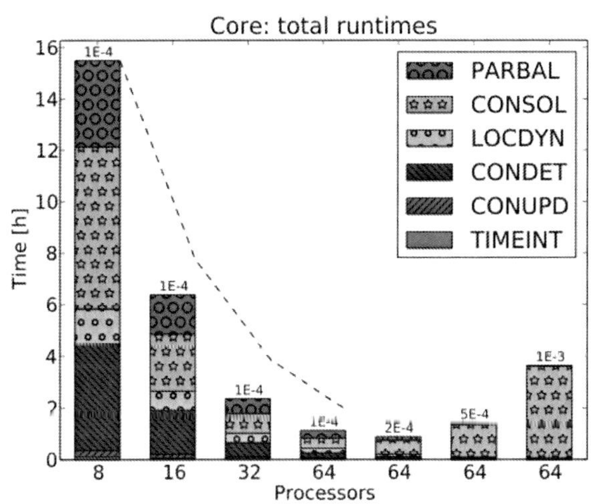

FIGURE 7: The simplified core example: total runtimes. TIMEINT stands for time integration. CONUPD stands for constraints update. CONDET stands for contact detection. LOCDYN stands for local dynamics assembling. CONSOL stands for the Gauss-Seidel constraints solution. PARBAL stands for parallel load balancing and communication not related to constraints solution. The dashed line indicates linear scaling for total runtimes.

Depending on the kinematic models, time step, constraints solver accuracy and number of processors, typical runtimes for a 10 second excitation of the 10x10x12 brick array can take from several hours to several days. Our general experience is that a fully implicit run of reasonable accuracy (in terms of the time step size and solver tolerances) can be made within 24 hours on 32 modern processors connect with fast Infiniband network.

It is perhaps worth noting, that SOLFEC includes also an improved penalty solver, allowing for larger time steps than in typical DEM codes. In this mode, no longer fully implicit, the runtime for the 10 second analysis of the 10x10x12 rigid body array is about 3.5 hours on 32 processors. Similar runtime can be achieved when using the implicit solver with liberally relaxed tolerances. It is the mater of future work to investigate the nature of qualitative differences between solutions obtained by means of the fully implicit and penalty based approaches.

CONCLUSIONS

In this paper, we have overviewed some aspects of SOLFEC that may be relevant in the context of modeling of the dynamics of AGR graphite cores. Three-dimensional modeling of multibody structures involving multiple impacts, shock waves, friction and fragmentation is an extremely challenging task. One aspect of the challenge is purely computational. We believe SOLFEC is well placed to tackle large scale problems. Another, more fundamental aspect of the challenge is to be able to capture the most significant physical phenomena that contribute to the modeled behavior. Validation of the computational framework implemented in SOLFEC is one of the essential elements of our ongoing and future work. Although SOLFEC implements some of the state of the art in multibody modeling, much more research is needed in order to better understand the mechanisms of energy dissipation and redistribution in large multibody, multi-impact structures. Since SOLFEC is an open research platform, with an increasing number of users and developers, we hope that we shall be able to keep in touch with the research out there, as well as contribute to it in some degree.

Acknowledgements

The support from EDF Energy is gratefully acknowledged. Some of the calculations were performed on a cluster of the Institute of Computer Modeling at the Krakow University of Technology.

References

Cundall, P. A. and Strack, O. D. L. (1979). A distinct element model for granular assemblies. *Geotechnique*, 29:47–65.

Jean, M. (1999). The non-smooth contact dynamics method. *Computer Methods in Applied Mechanics and Engineering*, 177, (3-4):235–257.

Koziara, T. and Bićanić, N. (2011). A distributed memory parallel multibody contact dynamics code. *International Journal for Numerical Methods in Engineering*, 87, (1-5):437–456.

Moreau, J. J. (1999). Numerical aspects of the sweeping process. *Computer Methods in Applied Mechanics and Engineering*, 177, (3-4):329–349.

Zhang, M. and Skeel, R.D. (1997). Cheap implicit symplectic integrators. *Applied Numerical Mathematics*, 25, 297–302(6).

Shaking Table Experimental Programme

Luiza Dihoru[1], Adam Crewe[1], Colin Taylor[1] and Tim Horgan[2]
[1] Dept. of Civil Engineering, University of Bristol, University Walk, Bristol, BS8 1TR, UK
[2] Atkins, The Hub, 500 Park Avenue, Aztec West, Bristol BS32 4RZ, UK
Email: Luiza.Dihoru@bristol.ac.uk

Abstract
The graphite components in AGR cores are subject to degradation processes that are predicted to lead to greater numbers of weakened and cracked components. The reactor core models used to assess the tolerability of the cores to seismic events need to represent higher levels of degradation. A shaking table programme has commenced, with the primary goal to provide experimental validation of the existing AGR core models. This paper presents the proposed rig work with main aspects of physical model design, rig manufacturing and dynamic testing being addressed. The candidate instrumentation systems and the relevant rig outputs are also described.

Keywords
Degraded core, seismic behaviour, physical modelling

INTRODUCTION

Due to the current position of the AGRs ('Advanced Gas-cooled Reactors') within their operating cycle, the safety case strategy has been updated to meet the concerns over the predicted age-related degradation of the cores. Therefore, it has become necessary for the reactor core models to represent higher levels of degradation of the graphite components and higher proportion of degraded core restraints than are currently considered. The existing dynamic models (GCORE, developed by Atkins and SOLFEC, developed by the University of Glasgow) may need to extend their capabilities to capture the behaviour of AGR cores with extensive component degradation.

It is recognized that seismic testing is needed to validate experimentally the aforementioned computer models and to enhance the understanding of core dynamics, especially where components may behave in ways not adequately modelled by computational analyses.

A programme of seismic testing has commenced at the University of Bristol (UOB) with the primary goal to provide experimental validation to GCORE and SOLFEC. The programme comprises the design, build and testing of a physical model of an AGR core which has increased potential for brick displacements due to component degradation or increased brick-to-brick clearances arising from advanced shrinkage. This paper presents the ongoing rig work for seismic testing and outlines the first decisions of model design, manufacturing and testing.

EXPERIMENTAL OBJECTIVES

The following aspects of damaged core behaviour have been identified as targets for experimental investigation. These aspects will provide the relevant inputs to the computer model validation work:

- Core distortion due to variation of horizontal direct clearance between bricks brought about by advanced shrinkage.
- Core distortion due to large percentage of singly and/or doubly axially cracked bricks.
- Core distortion due to displaced / locked / failed graphite keys.
- Core distortion from crack opening, even if adjacent keys are not damaged.

SCALE MODELS FOR SEISMIC TESTING

The model for seismic testing will be based upon the AGR cores of the reactors at Hinkley Point B (HPB) and Hunterston B (HNB) power stations at 34fpy ('full power years'). While the degree of representation of the model for the aforementioned AGR cores is important, the physical model design will be primarily guided by the objectives set out by the computer modellers for the validation work.

Components and Mechanisms to be Modelled

The complexity of the physical model will be determined by its number of components and level of precision in reproducing the geometry and scaled dimensions. The more complex the scale model, the higher its level of representation for the dynamic behaviour of a 'generic' HPB/HNB prototype. The simpler the physical model, the more sophisticated the mapping software that extrapolates the experimental results should be for a realistic prediction of the prototype response. UOB has embarked in a feasibility study that investigates several candidate rigs from the point of view of their feasibility and from the point of view of their experimental outcomes. Rig feasibility involves aspects such as manufacturing costs, manufacturing time, challenges of design and manufacturing, number of personnel required for handling and testing, complexity and convenience of the measurement system. The experimental outcomes refer to demonstration of core seismic tolerability for the damage scenarios listed in the previous section.

Table 1 shows possible combinations of components to be employed in the model for achieving certain experimental aims. Three types of model fuel bricks can be employed: uncracked, singly axially cracked and doubly axially cracked. The horizontal restraint system can be very simply modelled via a ring with various stiffness values or can be modelled in a more detailed way to emulate the restraint rod and restraint beam connections. It is important to note that for a fully representative rig, the boundary would need to simulate the dynamics of the core restraint and the response of any rings of bricks that are not explicitly modelled. For validation purposes only, the choice of boundary is less constrained - the computer model just needs to reflect the rig. Similarly, for a fully representative rig, the vertical restraint system would need to emulate the behaviour of the missing top and bottom layers of bricks and prevent sliding and unrealistic rocking of the bricks. It is recognized that adapting the restraint stiffness in the computer model is far simpler than trying to emulate physically such variation. Therefore, it has been decided that the physical model will not be fully representative. It will have a rigid restraint of constant and well-defined stiffness. According to the number of core components that are explicitly modelled, the rig can have various levels of complexity, but interpretation of results for real AGR core behaviour will require an additional and sophisticated process. The listed component combinations have been grouped around three possible candidate rigs as outlined in Table 1.

Variant No.	Components	Interactions/Mechanisms	Observations
1	One control channel rig Fuel bricks (uncracked, singly cracked and doubly axially cracked). Interstitial bricks Keys (loose, filler) Filler bricks Model restraint system	Interstitial channel distortion Key-keyway disengagement* Key locking* Key failure* Post-disengagement behaviour* Post-locking behaviour*	This arrangement is mainly to demonstrate the safe control rod insertion by monitoring the '3-point-angle' of the channel. *Modelling these aspects is limited by the insufficient number of components and oversimplified boundary conditions
2	One control channel with control rod rig Fuel bricks (uncracked, singly cracked and doubly axially cracked). Interstitial bricks Keys (loose, filler) Filler bricks Control rod Model restraint system	Interstitial channel insertion Key-keyway disengagement* Key locking* Key failure* Post-disengagement behaviour* Post-locking behaviour*	This arrangement is mainly to demonstrate the safe control rod insertion by modelling explicitly the control rods. *Modelling these aspects is limited by the insufficient number of components and oversimplified boundary conditions
3	Hybrid Array or Full Layer Fuel bricks (uncracked, singly cracked and doubly axially cracked). Interstitial bricks Keys (loose, filler) Filler bricks Model restraint system	Key-keyway disengagement Key locking Key failure Post-disengagement behaviour Post-locking behaviour Interstitial channel distortion Fuel channel distortion Effect of crack alignment on brick column stability and key system functionality. Variation of gaps between restraint system and graphite core during thermal cycling Variation of brick to brick gaps as a result of shrinkage.	Getting closer to the true prototype behaviour is dependent on the number of model components and the design of the boundary system. Control rods need not be explicitly modelled. Safe control rod insertion demonstrated via monitoring the '3-point-angle' of the channel. This rig can also investigate the fuel stringer '3-point-contact'.

TABLE 1: Examples of component combinations for physical modelling.

Scaling and Material Issues

In general physical modelling seeks an adequate approximation of the similitude relations between model and prototype. In this particular case, the following basic prototype and model facts have been considered:

- For HPB/HNB, a graphite density of 1.8 g/cm^3 has been historically assumed.
- For HPB/HNB, an unirradiated Young's modulus of 9.6 GPa has been assumed.
- The prototype Gilsocarbon compressive strength is about 80 MPa for HNB, and 70 MPa for HPB. The point contact collisions between components are considered rigid.
- Geometrical similitude required, including rocking feature of model fuel bricks, parallel walls for keys and dovetail shape for keyways.
- Dimensional precision required (*i.e.* tolerance of 0.1 mm or smaller for linear dimensions).
- Key-keyway clearances are scaled for the correct reproduction of rotational and translational movements of the key in the keyway (prototype clearance: 1.04 – 1.52 mm). The smallest practical scale for the model is a quarter-scale
- Mechanical properties of the model material should be stable with time, under normal environmental conditions.
- The ideal scaling factor for Young's modulus is 1. However, this poses severe restrictions on material properties and probably cannot be achieved in practice.

- Scaling factor for gravity is 1.
- Maximum dimension of model rig is dictated by the size of the UOB shaking table (3x3 m).
- Maximum weight of model rig is dictated by the capacity of the UOB shaking table (15 tonnes).

In principle, the models have to be easy to manipulate and to monitor. The larger the scale of the model, the more time is needed to assemble and more effort is required in handling. On the other hand, in smaller scale models the measurements are more difficult to make (*i.e.* limited space for transducers) and the signal / error ratio becomes smaller. It is also important to note that the dynamic behaviour of large arrays may be significantly different to that of smaller ones and that a higher degree of sophistication would be required in computer model validations for smaller arrays. Overall, the above arguments are driving the physical modelling choice towards the models with a larger number of components, but relaxing the requirement for full AGR core representation.

A summary of scaling factors for earthquake response of structures can be found in Crewe (1998). In general, a true replica model implies simultaneous duplication of inertial, gravitational and restoring forces and full compliance with the similitude laws. Such a model would require scaling of density and stiffness at the same time. Finding a material whose properties satisfy the scaling requirements simultaneously is practically impossible; therefore, an adequate approximation has to be sought. Another method employed in physical modelling is the artificial mass simulation method. It implies the presence of additional material of a non-structural nature to simulate the required density of the model. Such mass can be lumped or distributed. This method is difficult to apply to the scaled AGR core model because of the large number of components that have a role in system's dynamics. Distributing an artificial mass within such a complex array of rigid blocks would be technologically impractical. The third type of modelling applies to cases where gravity forces can be neglected. In the particular case of a graphite core under seismic loading, the gravitational forces cannot be neglected; therefore, using the third type of scaling law is out of the question. It is therefore proposed that the graphite core model should be an 'adequate model' which maintains 'first-order' similarity. 'First-order' similarity implies that the physical parameters with significant influence on the seismic response are accurately scaled, while the 'second-order' parameters are only be approximately scaled. In this way a modified version of a true replica model will be created. For this research programme, it is proposed that the geometrical properties of the core will be scaled, as the channel shapes and the general distortion of the core are governed by the brick-to-brick and the key-keyway clearances. It has been decided that the brick and key design will be a quarter scale design based on the HPB/HNB prototype. All the clearances in the model will be quarter scaled. The scaling of material properties has also to consider the dynamic problem that is at the centre of this investigation. During a seismic event, the core will behave as an array of rigid bodies in which the relevant forces are the impact forces generated during the collisions between the components. As impact forces depend on the local contact properties (*i.e.* contact stiffness and coefficient of restitution), then the Young's modulus of the component material becomes relevant in scaling. It is important to observe, that the contact properties are different for the normal and for the shear contact. The energy restitution after a brick-to-brick collision depends heavily on the actual layout of components in a zone of investigation (*i.e.* presence or absence of bearing key, presence or absence of integer key, locking of key, *etc*). Brick-to-brick testing of prototype model components will be carried out at UOB to determine the values of contact properties for various component combinations. The selection of the model material will have to seek a reasonable approximation for the ratio between the scaling factor for density (S_ρ) and the scaling factor for stiffness (S_E) of the model component. This ratio is described by the basic scaling law in the following equation:

$$S_\rho = S_E / S_L \qquad [1]$$

where S_L is the scaling factor for length. For a quarter scale model, Equation [1] becomes:

$$S_\rho = 4 \times S_E \qquad [2]$$

Table 2 presents the results of Equation [2] for a number of candidate materials.

Material	Density(kg/m^3)	Young's Modulus (GPa)	Compressive Strength (MPa)	S_r	S_E	S_r / S_E
Commercial Graphite	1800	9.60	70	1	1	1
Nylon 12	1020	1.80	75	0.57	0.19	3.02
POM(Acetal)	1410	2.70	90	0.78	0.28	2.79
Reinforced Acetal	1580	9.00	100	0.88	0.94	0.94
PPS(Fortron)	1600	13.00	93	0.89	1.35	0.66
LCP(Vectra)	1610	13.00	90	0.89	1.35	0.66
Alluminium Alloy	2700	70.00	110	1.50	7.29	0.21

TABLE 2: Equation [2] for candidate model materials. Note: Property values are indicative. PPS (Fortron®) is polypheniline sulphate with 40% glass reinforcement. LCP(Vectra®) is a liquid cristal polymer with 30% glass reinforcement (supplier: Ticona).

Among the candidate materials, the engineering thermoplastic AF907 Acetal® (polyoxymethylene or POM) has been selected for its reasonable density/stiffness ratio, but also for its high rigidity which makes it suitable for precision machining (dimensional tolerances down to 0.1 mm). Also, this material exhibits excellent dimensional stability thanks to its non-hygroscopic behaviour.

The friction between the surfaces in contact within the array is another property of interest. Table 3 presents the static and the dynamic coefficient of friction for two materials in contact. It can be noted that the model components made of POM (Acetal) exhibit comparable friction coefficients with the graphite bricks at reactor operating temperature.

Proposed Models for Seismic Testing

The physical models for seismic testing have been decided based on the scaling and material considerations presented above, on the capacity and size of the shaking table and on previous experimental experience with the measured displacements in existing array rigs. The smaller and stiffer the array, the smaller the relative displacements between the components, making measurement more challenging. Recent measurements on mini-core tests (Roscow *et al.*, 2010) show that the displacements in a 10 x 10 section are ~10 mm maximum for the 100% cracked core, but only ~5 mm for the 50% cracked core, and as small as ~1 mm for the intact core. It is also important to note, that smaller and stiffer models are likely to respond at higher frequencies and that such behaviour would be a departure from the actual prototype for which a 'natural frequency' of 2-3 Hz has been historically assumed. Lowering the 'natural frequency' of the model can be done via increased slackness in the system. As the horizontal clearance value should be representative for the 34 fpy (*i.e.* it has a well-defined scaled value), then slackness can be increased via increased key-keyway clearances and / or via increased area section of the model. If the key-keyway clearances are to be kept quarter-scale of the prototype value, then the only route for increasing slackness is via a larger model section.

Two model configurations are proposed:

1) A full core single layer rig (quarter scale, 24 fuel bricks across the diameter, array diameter 2.76 m, array height 0.21 m, array mass 1200 kg)

Its restraint system will of simplified design and completely rigid. The base of the rig will be frictionless to allow sliding. This rig will provide boundary margins of displacements for the components under dynamic excitation. Its main role will be to provide proof of concept and possibly highlight experimental aspects that need considering in the second phase, when a multilayered rig will be tested. The single layer rig is a necessary learning stage for set-up, test and measurement, as well as a tool that provides a better feel for response.

2) A multilayered full or near-full core rig (quarter scale, 20 bricks across the diameter, 4 layers, array diameter 2.30 m, array height 0.84 m, array mass 4100 kg). The rig can be expanded to accommodate 6 or 8 layers if requested.

Its restraint system will of simplified design and completely rigid. The base of the rig will be a seating grid that allows only rocking. More than one such grids can be used to emulate several values of horizontal direct clearance between bricks.

Both models will be manufactured from stock shapes of POM (Acetal), via a 3-phase manufacturing process which consists of extrusion, annealing and precision machining. The expected tolerances to the linear dimensions are between 0.05-0.1 mm.

No.	Material 1	Material 2	m_{static}	$m_{dynamic}$
1	POM (Acetal)	Steel	0.14	0.21
2	POM (Acetal)	POM (Acetal)	0.19	0.15
3	POM (Acetal)	PA (Polyamide)	0.04	0.06
4	POM+20% PTFE	PA + 20% PTFE	0.03	0.04
5	Graphite (temp 20 deg C)	Graphite (temp 20 deg C)	0.10	0.10
6	Graphite (temp 350 deg C)	Graphite (temp 350 deg C)	0.20	0.20
7	Aluminium (dry)	Aluminium (dry)	1.05	1.40
8	Aluminium (greasy)	Aluminium (greasy)	0.30	0.32

TABLE 3 Friction coefficients for selected pairs of materials.

DYNAMIC LOADING

The dynamic testing programme will be designed to suit the validation needs of GCORE and SOLFEC. The earthquake simulator at UOB is capable to generate a wide range of inputs, such as: sinusoidal waveforms, sinusoidal sweeps, real earthquakes and artificial earthquakes. For this particular application, the following types of inputs will be considered:

- Sine dwell inputs to explore the frequency response of the model and its ability to replicate the basic mechanics. The sine dwell probes will reveal onset of motions more clearly than would more random seismic loads.

- Sine sweep inputs to explore the frequency response of the model and to investigate its response over a wider range of frequencies. Degradation of stiffness (arising from key disengagement for example) can be observed via repeated sine sweeping.

- Seismic inputs run at different amplitudes so as to detect onset of changes in behaviour. Running sets where the dominant input frequency is below, at, and above the 'natural frequency' of the core may also be considered with the purpose of checking the frequency response of the numerical models. The input records will be based on power spectra and not on response spectra, as the former will contain the same input energy, therefore the tests will have the same base for comparison.

Real time acceleration and displacement of selected components inside the array will be measured. Time histories and frequency responses as well as phase information (ideally displacement vs. velocity) will be the target test outputs. Since the characterisation of the starting conditions for the physical core model will be difficult, making direct validation comparisons will be difficult. The comparisons with the numerical models will have to be more generic, perhaps looking at envelopes of response. It is recognised that the existing models (GCORE and SOLFEC) will have to be enhanced in order to interpret or extrapolate and map the observed physical model behaviour. The complexity of their mapping capabilities will be inversely proportional to the complexity of the experimental rig.

MEASUREMENT SYSTEMS

Two types of measurement will be made: acceleration and displacement. SETRA 141A capacitive-type accelerometers will be used. The accelerometers generally have a calibrated range of +/-8g that is traceable to national standards. In each case the calibration applies from DC to well above 100 Hz. The displacement measurement system will record the real-time tracking of the centres of selected fuel and interstitial bricks. The aim is to track all of the bricks across the 2.76 m diameter, at a resolution of 0.1 mm and input frequencies up to 50 Hz. For the multilayer array, the aim is to track the centres of the interstitial bricks forming a channel and such tracking to be performed for 5-10 channels at a time. The exploratory work for determining the suitable systems for this task is ongoing. Certain high-speed, high-resolution vision systems and several non-contact displacement transducers are currently being assessed. Trial tests on a 10x10 array have been conducted on the shaking table for this purpose. Figure 1 shows a photo of the 10x10 array with reflective markers (M1-M5) attached to several of its bricks as part of the vision system measurement procedure.

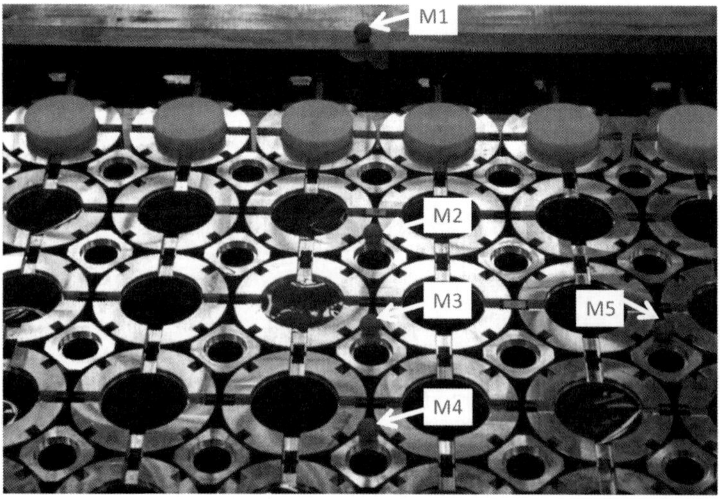

FIGURE 1: Trial array (10x10) with vision system markers (M1 to M5).

CONCLUSIONS

An ongoing experimental programme to support the validation of existing AGR core computer models has been presented. The seismic rig will be a quarter scale model that complies with the geometrical similitude laws and will present sufficient versatility of design to be tested in two geometrical layouts (*i.e.* full core layer and multilayer array). The stiffness of its restraint system will be simulated numerically. The rig will be able to represent the 'mechanisms' that challenge the core functionality (*i.e.* core distortion due to cracking, key-keyway disengagement, key jamming, key failure and variations of brick to brick gaps). The rig will be able to emulate both frictionless and anti-sliding base conditions. Displacement and acceleration of individual bricks and fuel and interstitial channel shapes will be the main rig outputs to be further employed in model validation work.

Acknowledgements
The authors would like to thank EDF Energy for financial support and permission to present this paper. The views expressed in this paper are those of the authors and do not necessarily represent those of EDF Energy.

References

Crewe, A. J. (1998). The characterisation and optimisation of earthquake shaking table performance. PhD Thesis, University of Bristol.

Roscow, A. R., Skelton, J. and McLachlan, N. (2010). Continuing development of the quarter scale core model to support the safe performance of AGR cores, *Securing the Safe Performance of Graphite Reactor Cores*, Conference Proceedings, Univ. of Nottingham, United Kingdom, 24-28 November 2008, 203-209.

A Novel Approach for Evaluating Surface Deposits on the Channel Wall Trepanned Graphite Samples

Peter J. Heard[1], Malcolm R. Wootton[2] and Peter E. J. Flewitt[1,3]

[1]Interface Analysis Centre, Oldbury House, University of Bristol, Bristol BS2 8BS, UK

[2] Magnox Ltd., Oldbury Technical Centre, Oldbury Naite, Gloucestershire BS35 1RQ, UK

[3] School of Physics, HH Wills Laboratory, University of Bristol, Bristol BS8 1TL, UK

Email: p.j.heard@bristol.ac.uk

Abstract

Samples have been trepanned from the fuel and interstitial channel walls of PGA graphite reactor cores of two operating Magnox gas cooled power stations after a period of service. These samples have been considered explicitly for the presence of deposits on the channel facing surfaces. A combination of focused ion beam milling and imaging has been used to determine the presence of such deposits and where present to make measurements of the thickness. These thicknesses vary from a few nanometres to tens of micrometres. In addition, both the chemical composition and chemical state have been investigated using energy dispersive X-ray microanalysis in a scanning electron microscope and Raman spectroscopy respectively. The results are discussed by comparing the microstructure, composition, chemical state and origins of the deposits with the parent graphite.

Keywords

Raman, focused ion beam, EDX.

INTRODUCTION

Pile Grade A (PGA) polygranular graphite is used as a moderator in Magnox reactors. The graphite bricks are manufactured by extrusion of a mixture of needle coke filler particles with coke flour and coal pitch binder, as described previously (Kelly, 2000, Hodgins, 2010, Heard, 2010a). Complex interconnected nano- and micro-scale pathways have been observed to link many of these pores throughout the microstructure (Hughes, 2008). The polygranular PGA microstructure, and mechanical and physical properties, changes over the period of exposure to the service environment of hot CO_2 gas and neutron irradiation (Best, 1985). These changes arise from two processes: (1) neutron irradiation, which introduces point defects into the crystal lattice; and, (2) radiolytic oxidation by the CO_2 gas as a result of gamma irradiation produced by the nuclear fission. The latter cause graphite mass loss and thus a decrease in strength. In addition, the reaction can lead to instabilities in the CO_2-graphite interaction that can lead to the formation of carbonaceous solid redeposition on various surfaces within the gas coolant circuit (Evans, 1982). Periodically, cylindrical test samples are trepanned from the fuel and interstitial channels to monitor changes in the amount of porosity and weight loss (Heard, 2010a). In the course of these examinations, a 5-10 μm thick deposit was observed on the fuel channel wall of a specimen obtained from one of the power stations.

MATERIALS AND EXPERIMENTAL TECHNIQUES

Virgin PGA graphite supplied by Magnox Electric Ltd was part of the archived material from the construction of the reactor cores at a Magnox power station (identified as station A). Specimens of the same dimensions as those radiolytically oxidised were prepared for analysis as detailed previously (Heard, 2010b). Eight radiolytically oxidised graphite specimens were extracted by trepanning from the interstitial and fuel channel reactor core bricks from two Magnox power stations. These were cut to produce cylindrical specimens as detailed below. The sample identifications, weight loss due to radiolytic oxidation (%) and location within the reactor cores are presented for both power stations, code A and B, Table 1. Specimens A15 and B37.5 were mid-section specimens, and therefore did not contain a channel wall surface whereas the remaining six specimens contained channel wall surfaces. Specimens are identified according to the power station of origin (A or B), the channel wall surface (interstitial, I, or fuel, F) and the percentage weight loss.

Sample reactor A	**AI7.5**	**A15**	**AI16.3**	**AF15.1**
Weight loss (%)	7.5	15	16.3	15.1
Location	Interstitial	Cut surfaces.	Interstitial	Fuel channel wall.
Sample reactor B	**BI31.7**	**BF30.9**	**BF17**	**B37.5**
Weight loss (%)	31.7	30.9	17	37.5
Location	Interstitial	Fuel channel	Fuel channel	Cut surfaces.

TABLE 1: Radiolytically oxidised specimens.

The focused ion beam FEI Strata FIB-201 workstation used for milling and imaging the specimens has been described previously (Hughes, 2008). A beam energy of 30 keV was selected, with the beam current adjusted according to milling and imaging requirements, in the range 12 nA – 1 pA. Trenches were cut into the material at higher beam currents, starting at 12 nA and reducing to 2.7 nA or 1 nA for final preparation. A Zeiss EVO MA10 scanning electron microscope operating at 30 keV was used to obtain surface images of the specimens. Images and spectra were obtained with a beam current of 300 pA in secondary electron mode. The instrument was equipped with a Princeton Gamma-Tech OS16-P008 energy dispersive X-ray detector with WinEDS software for quantitative chemical analysis. Raman spectra were obtained using a Renishaw Raman System 2000 instrument. Specimens were mounted under a Leica optical microscope and excited with 514 nm wavelength argon ion laser light. A 50x objective lens was used on the system to maximise the signal obtained. Spectra were acquired in the Raman shift range 1000 to 3000 cm^{-1}, using acquisition times of 360 s per spectrum. The 520 cm^{-1} peak from a silicon specimen was used for calibration.

RESULTS

Fuel channel wall specimens BF17 and BF30.9 were taken from reactor core bricks at power station B. Figures 1(a) and (b) show FIB images from BF30.9 where the appearance of the surfaces differed considerably from that of radiolytically oxidised material from within the brick (Figure 1(c)). The surfaces here resembled a cauliflower; an agglomeration of irregular spheres of material of up to 2 μm in diameter. These spheres had relatively smooth surfaces on the sub-micrometre scale, especially for the material within the various cavities that existed across the surface. Protruding material contained smaller spheres, and hence a rougher appearance. FIB sections from BF17 and BF30.9, Figure 1(b), showed that very little porosity existed within the material. The limited porosity within 10 μm of the surface appeared to arise from cavities between spherical particles that had partially grown together. The amount of porosity was very low compared to that of radiolytically oxidised material from the centre of the brick, in which large pores were present due to the preferential oxidation of matrix material. Furthermore, features such as Mrozowski cracks and lamellar graphite were absent at the channel surface. During FIB milling it was noted that the surface material etched more quickly than either virgin or radiolytically oxidised graphite. Hence, it is proposed that the surface material is a deposit, rather than original graphite, although there was no contrast between the two materials in the FIB images. At depths greater than about 10 μm on the BF30.9 specimen, pores resembling Mrozowski cracks and ligaments were observed between lamellar structures (marked by arrows in Figure 1(b)), indicating that at this depth, an original graphite filler particle had been reached. A thin layer of fine porosity appeared to form a boundary between the deposit, characterised by spherical particles, and filler graphite. This would imply a deposit thickness of ~7 μm at this position. Further sections showed a less clear deposit depth, and spherical particles sometimes existed 20-30 μm beneath the surface. If the deposit had filled the pores normally associated with oxidation of the material, this would be expected.

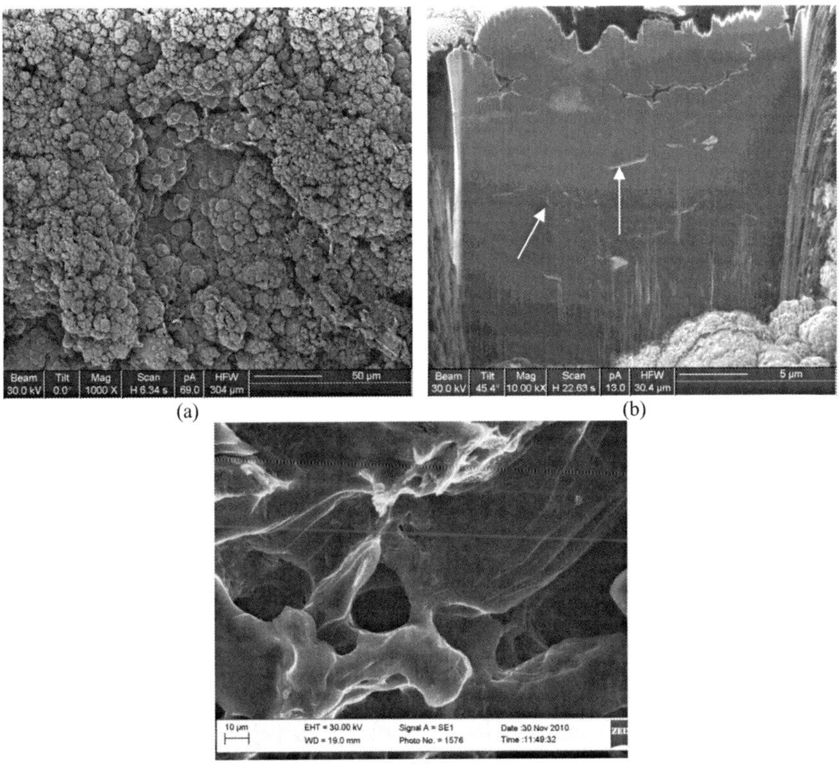

FIGURE 1: (a) FIB image of the surface of BF30.9 fuel channel wall; (b) FIB section of BF30.9 fuel channel wall where arrows show ligaments and Mrozowski cracks characteristic of graphite filler particles; and, (c) SEM image from the reverse side of BF30.9 fuel channel wall showing porosity associated with radiolytic oxidation.

At the FIB sites excavated, the BF17 specimen was covered with deposit at least 20 μm thick. No fine porous layer was observed that would have indicated an interface with parent graphite. A scraped region on the specimen was also sectioned where the surface deposit had apparently been removed by mechanical disturbance (of unknown provenance). Sections through this region show lamellar graphite and Mrozowski cracking normally associated with graphite filler particles, as well as the presence of needle graphite in some regions. Some of the pores beneath the scraped region showed signs of erosion indicating radiolytic oxidation. The depth of the scraped region on this specimen was estimated from the FIB images to be approximately 20 μm, and therefore the thickness of the deposit on the remainder of the specimen is assumed to have this value.

The reverse sides of the BF17 and BF30.9 specimens were examined. These showed the skeletal structure typical of radiolytically oxidised PGA graphite, as evident from Figure 1(c) (Heard, 2010b). This microstructure is very different from that associated with the surface deposits described above.

EDX analysis was performed on the channel wall faces of BF17 and BF30.9 as well as the reverse side of BF30.9. Figure 2 (upper trace) shows the spectrum obtained from the front, deposit face of BF30.9. The majority of the material is carbon, with smaller peaks observed corresponding to oxygen, iron and manganese (the vertical scale is logarithmic). Additional peaks assigned to copper and zinc are from the brass specimen holder. Standardless quantification using the WinEDS software gave the iron content to be 0.03 at%. The small amounts of other elements detected, including silicon, sulphur, potassium, calcium and titanium were not always present at other locations. Figure 2 (lower trace) shows a spectrum from the reverse side of BF30.9. Iron was not observed here, but small peaks from sodium, chlorine and silicon were present. For comparison, EDX spectra obtained from the surface of virgin PGA showed only carbon, oxygen and the copper and zinc peaks from the specimen holder.

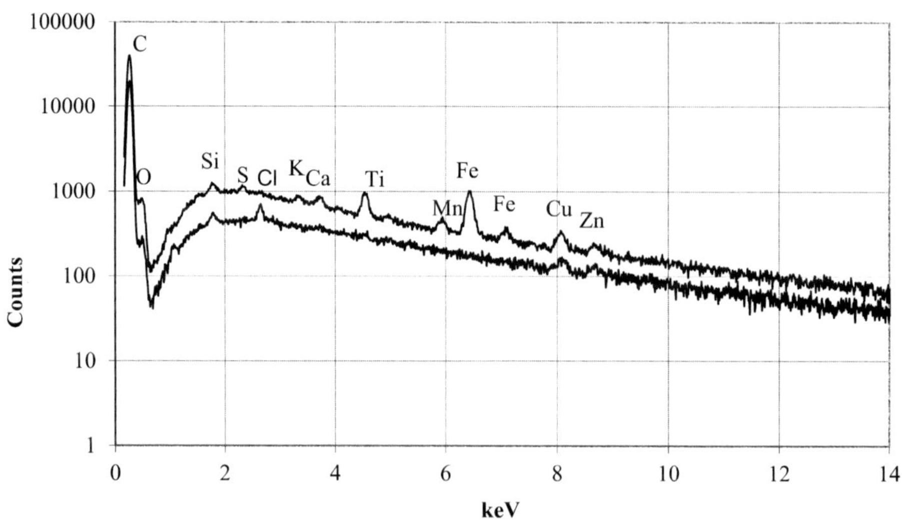

FIGURE 2: EDX spectra from BF30.9 fuel channel wall (upper trace) and the reverse side (lower trace).

Specimen AF15.1 was taken from the fuel channel wall of power station A. A FIB image of the surface of the specimen, Figure 3(a), revealed the presence of smooth regions, up to several hundred micrometres in diameter, interspersed with rougher regions. A FIB section, Figure 3(b), showed the presence of Mrozowski cracks. The large, smooth regions are consistent with the presence of filler particles in the graphite. The well-defined layer of fine porosity found in the power station B specimens was absent. If a small amount of deposit is present in this specimen, it is of minimal thickness and limited in spatial extent compared with the power station B specimens. The reverse side of the above specimen was not examined, but a power station A specimen of similar weight loss (A15) has been observed previously (Heard, 2010b). This specimen was approximately 12-18 mm from the fuel channel wall. FIB images of this specimen showed a skeletal structure (mixture of pores and ligaments) typical of radiolytically oxidised graphite. An EDX spectrum from the AF15.1 specimen was similar to Figure 2 (upper trace) and showed mainly carbon with small amounts of oxygen, iron, manganese, silicon, sulphur and calcium.

(a) (b)

FIGURE 3: FIB images from (a) the surface; and, (b) a section of specimen AF15.1 fuel channel wall.

Interstitial channel wall specimen BI31.7 was taken from power station B. A FIB image of the surface of the specimen is shown in Figure 4(a). The surface resembled that of Figure 1(a), an agglomeration of deposited near-spherical features. A FIB section of the specimen, Figure 4(b), showed the deposit to be separated from the main body material by a well-defined finely porous layer. At a depth of approximately 15 μm, a feature resembling a Mrozowski crack was present. The thickness of material above the porous layer was approximately 3 to 5 μm. An EDX spectrum from BI31.7 showed, in addition to the major carbon peak, sodium and chlorine (about 1 at%). Iron was present (0.04 at%), although spectra acquired from cavities within the specimen contained less iron. Sulphur, potassium and calcium were also present at various locations across the specimen.

Specimens AI7.5 and AI16.3 were taken from the interstitial channel walls of power station A. FIB images showed features that were recognisably graphite, containing structures such as lamellae and needle graphite. However, much of the surface appeared to be coated with a very thin layer of small particulate material. FIB sections showed the particles to be submicrometre in diameter, while the underlying graphite comprised the usual lamellar features and Mrozowski cracks associated with PGA graphite. An EDX spectrum for specimen AI7.5 showed mainly carbon with small quantities of iron, silicon and calcium. The majority of the spectra taken from the surface of specimen AI16.3 showed a small iron content with additional peaks attributed to silicon, sulphur and calcium.

Raman spectra taken from five points on the virgin PGA graphite specimen, Figure 5(a), are shown on the same graph, but displaced by 5000 counts each vertically for clarity. The spectra were similar from each point visited on the specimen, and exhibited several strong peaks in the range 1000 to 3000 cm^{-1}. Spectra taken from the reverse side of specimen

BF30.9, Figure 5(b), are also displaced by 5000 counts vertically for clarity. This side of the specimen did not have a deposit, and would be expected to be typical PGA graphite, although radiolytic oxidation and neutron damage would have occurred. The spectra showed peaks similar to those in the virgin PGA graphite, Figure 5(a), although they were broad and the height of the D^* peak was lower. Spectra taken from the deposit at power station B fuel channel wall (BF30.9), Figure 5(c), were consistent across the specimens and exhibited the graphite D and E_{2g} (G) peaks but with further broadening and with a high background signal across the wavelength range. There was no evidence for a D^* peak; rather, an increase in the background of these spectra over the corresponding range of wavenumbers. Spectra from power station A were similar for the reverse side, and for the small amounts of local deposit, to power station B.

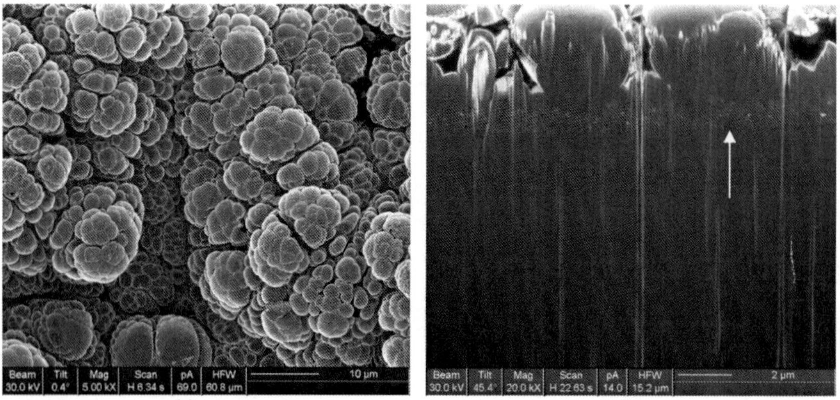

(a) (b)

FIGURE 4: FIB images from (a) the surface of BI31.7 interstitial channel wall; and, (b) section at high magnification showing line of porosity (arrow).

DISCUSSION AND CONCLUSIONS

The channel wall surfaces of the graphite specimens obtained from the B reactor core were very different from the bulk. Surface images revealed a cauliflower-like appearance; an agglomeration of irregular spheres of 2 μm diameter or below. FIB sections through the material revealed limited porosity compared with that of radiolytically oxidised graphite. Frequently a line of fine scale pores were observed that corresponded with the transition between the surface deposit and the substrate parent PGA graphite. However, this did not offer an unambiguous identification of the deposit depth. Further into the material however, at depths of 20 μm or more, FIB sections revealed the presence of thin micro-cracks, or Mrozowski cracks that are normally present in graphite filler particles. In the B station fuel channel 30.9% weight loss material (BF30.9), there was a well-defined layer of fine porosity (pores of 100 nm diameter or less) of approximately 500nm thickness beneath the surface of the specimen at a depth of 5 to 20 μm. It was notable that the crack-like features normally associated with the filler graphite particles only appeared beneath this layer, while the porosity above the layer was that associated with the gaps between the deposit spheres.

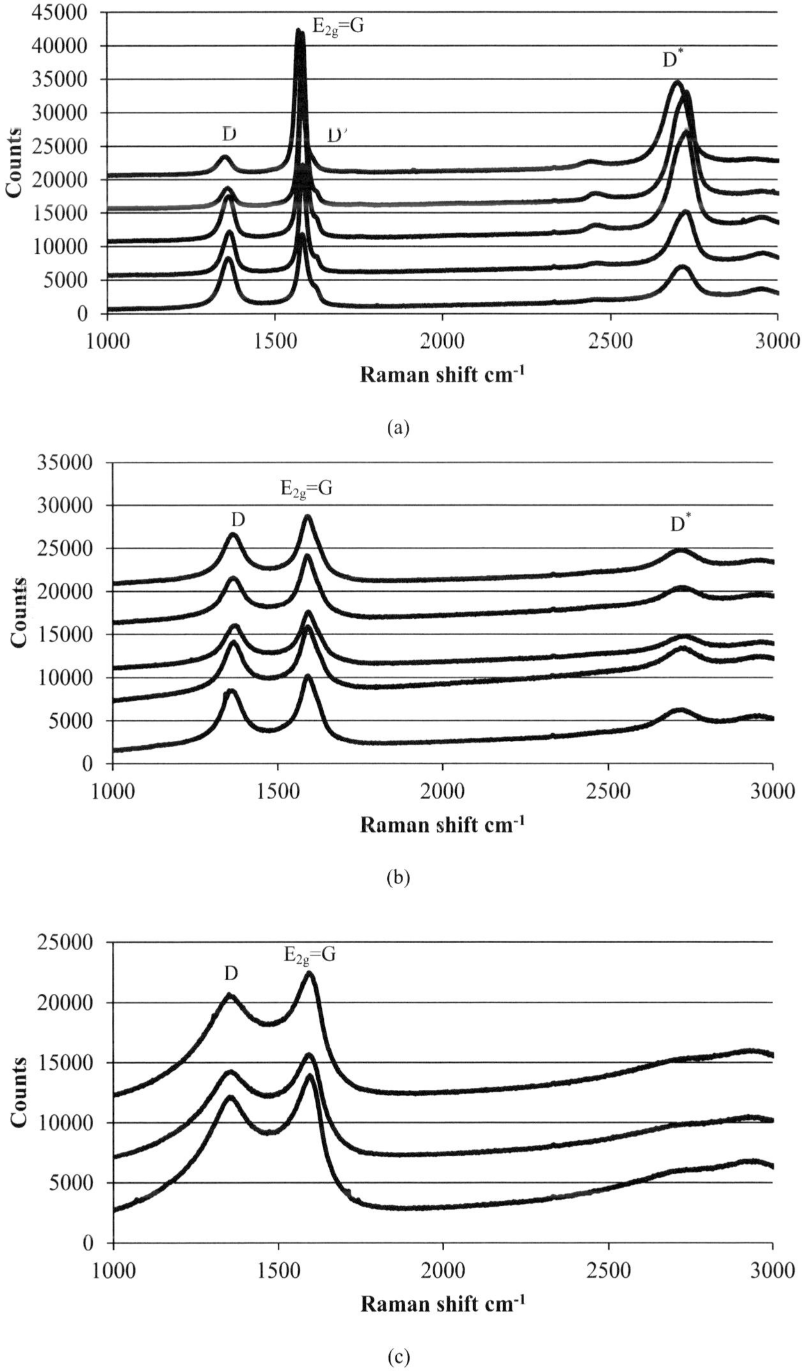

(c)

FIGURE 5: Raman spectra from (a) five points on the virgin PGA graphite specimen showing the E_{2g} (G) vibrational mode peak, the D and D' disorder peaks and the second order D^* peak; (b) five points on the reverse side of specimen BF30.9; and, (c) three points on specimen BF30.9.

Raman spectroscopy has been recognised to be a valuable technique for examining the different types of graphite that can form, such as single crystal graphite through to carbon black (Nikiel, 1993). However it has proved difficult to reconcile the exact form and origin of the various spectra because of the sensitivity of the spectra to both the crystalline state of the graphite and the impurity content. Certainly the first order Raman spectra obtained from single crystal graphite show a strong peak at 1580 cm^{-1} (G). This peak has been assigned to the in-plane E_{2g} (G) zone centre mode. By comparison, micro-crystalline graphite and disordered carbon introduce additional peaks at 1330 cm^{-1} (D) and 1620 cm^{-1} (D'). In addition, Elman (1981) reported sharp peaks at 2327 cm^{-1} and 2435 cm^{-1} and a broad peak at about 2468 cm^{-1} in single crystal spectra. These peaks are several orders less intense than the second order peaks at about 2700 cm^{-1} (D*). As a consequence, since the virgin PGA spectra in Figure 5(a) will be derived from single crystal filler particles and a matrix comprising graphite particle flour and graphitised pitch, it is to be expected that a strong G peak will originate from the former, and the latter will give rise to D, G and D* peaks. The variation in the relative intensity of the spectra will be a consequence of (1) the region sampled and therefore the relative proportion of filler particle and matrix sampled and (2) the size of the crystallites. It has been proposed by Tuinstra and Koenig (1970) that the relative intensity of the D and G peaks will be a function of crystallite size. In general, the spectra obtained from the reverse side of specimens A and B following radiolytic oxidation had characteristics similar to the virgin graphite. However, there was degradation of the spectra with increase of D and G peak widths and changes in the relative intensity. This is probably a consequence of irradiation damage introduced into the crystalline graphite in the form of point defects followed by preferential loss of matrix graphite; in particular the graphitised pitch phase. The spectrum obtained from the redeposit surface graphite, specimen BF30.9, Figure 5(c), had characteristics described by Tuinstra and Koenig (1970) where the D and G peaks broaden and the relative intensity of the D peak increases as the amount of unorganised carbon increases and the size of the crystals decreases. It is noteworthy that the second order D* peak at about 2700 cm^{-1} also decreased in intensity and there was a general broad intensity rise peaking at about 3000 cm^{-1}. At present, this provides an indication of differences between the deposit and the parent material, but there is no explanation for the origin of this difference.

Both the elemental composition analyses and the Raman spectroscopy provided positive support for differences between the surface deposits and the parent underlying irradiated PGA graphite. Raman spectra from the channel wall surfaces differed from those obtained from within the bricks and from the virgin PGA. In particular, the ratio of the D to E_{2g} Raman peaks were higher in the deposit material, the valley between the peaks was lower, the D^* peak was absent, and the general background levels were higher. For virgin PGA, the ratio of the D/E_{2g} peak heights ranged from 0.10 to 0.67 across the specimen, and these figures would correspond to graphite crystallite sizes from 44 nm to 6.5 nm, respectively (Tuinstra, 1970; Theodosiou, 2010). For the deposit material, the ratio ranged from 0.93 to 0.97 giving crystallite sizes of approximately 4.5 nm.

Taken together, these data pointed to the presence of a deposit on the channel walls (both fuel and interstitial) of the power station B specimens of between 5 and 20 μm thickness. A thin layer of fine porosity formed an interface between the deposit and the underlying graphite in some regions, see Figure 4(b), but in other regions this was not observed. One explanation for this could be that the deposit extended within the pores of the material, beyond the maximum FIB milled depth (about 50 μm). Elemental and chemical state measurements showed the layer to be predominantly carbon, either amorphous or with very small crystallite size. No contrast was found in the FIB images between the deposit and the underlying graphite and so it was not possible to determine the morphology of the graphite in the absence of a deposit to establish whether the graphite had been either oxidised or protected from oxidation due to the deposit. The deposit thickness did not appear to correlate with the specimen weight loss; the thickness on the BF17 specimen was of a similar thickness to that on the BF30.9 specimen.

The situation regarding the power station A material was not quite as clear. Similar globular features were observed in some regions on the channel walls, but FIB sections generally revealed graphite filler with its usual crack-like features, close to the surface. EDX analysis again showed the presence of iron and sometimes other elements at the surface. The Raman spectra varied across the specimen, showing deposit-like spectra in some positions, and graphite-like spectra in others. The sampling depth of Raman is approximately 0.3 μm, and hence it is possible that deposit-like spectra could be obtained from quite thin layers of material. From these data, it is evident that some deposit exists on the power station A channel walls, but that it is patchy and thin and therefore insignificant compared with that found on the power station B specimens.

Acknowledgements

The authors would like to thank Magnox Ltd for funding this work and permission to publish this work.

References

Best, J. V., Stephen, W. J. and Wickham, A. J. (1985). Radiolytic graphite oxidation. *Progress in Nuclear Engineering,* 16, 127-178.

Elman, B. S., Dresselhaus, M. S., Dresselhaus, G., Maby, E. W. and Mazurek, H. (1981). Raman-scattering from ion-implanted graphite. *Physical Review B,* 24, [2], 1027-1034.

Evans, H. E., *et al.* (1982). Carbonaceous-solid formation in Magnox reactor coolant. CEGB Report TPRD/B/0133/N82.

Heard, P. J., Wootton, M. R., Moskovic, R. and Flewitt, P. E. J. (2010a). Crack initiation and propagation in Pile Grade A (PGA) reactor core graphite under a range of loading conditions. *Journal of Nuclear Materials,* 401, 71-77

Heard, P. J., Flewitt, P. E. J., Wootton, M. R. and Moskovic, R. (2010b). A comparison of the cracking behaviour of virgin and radiolytically oxidised PGA graphite. University of Bristol Interface Analysis Centre Report PJH/IAC/02/10/C01.

Hodgkins, A., Marrow, T. J., Wootton, M. R., Moskovic, R. and Flewitt, P. E. J. (2010). Fracture behaviour of radiolytically oxidised reactor core graphites: a view. *Materials Science Technology,* 26, [8], 899-907.

Hughes, G. M., Heard, P. J. and Wootton, M. R. (2008). Examination of the pore structure of virgin PGA graphite using focused ion beam microscopy and Amira 3-D reconstruction software. University of Bristol Interface Analysis Centre Report GMH/IAC/06/07/C01.

Kelly, B.T. (2000). The structure and manufacture of nuclear graphite. In:- Irradiation Damage in Graphite Due to Fast Neutrons in Fission and Fusion Systems, IAEA-TECDOC-1154, IAEA.

Nikiel, L. and Jagodzinski, P. W. (1993). Raman-spectroscopic characterization of the oxidation of nuclear-grade graphites. *Applied Spectroscopy,* 47, [12], 2087-2092.

Theodosiou, A., Carley, A. F. and Taylor, S. H. (2010). Ion-induced damage in graphite: a Raman study. *Journal of Nuclear Materials,* 403, 108-112.

Tuinstra, F. and Koenig, J. L. (1970). Raman spectrum of graphite. *Journal of Chemical Physics,* 53, [3], 1126-1130.

^{3}H and ^{14}C Release in UK Nuclear Graphite by Chemical Treatment

Lorraine McDermott, Abbie N. Jones and Barry J. Marsden
Nuclear Graphite Research Group, School of MACE, The University of Manchester, Sackville Street, M13 9PL
Email: lorraine.mcdermott@manchester.ac.uk

Abstract
Irradiated nuclear graphite waste may account for up to 30% of any future UK geological ILW disposal facility (NDA, 2002). In order to make informed decisions of how best to dispose of large volumes of irradiated graphite waste within the UK nuclear programme, it is necessary to understand the nature and migration of isotopes present within the graphite and ultimately how these short and long term isotopes may behave prior to and during disposal. Various proposed decontamination and immobilisation treatments, within the EU Euroatom FP7 CARBOWASTE programme have been explored (von Lensa, 2008; von Lensa *et al.*, 2011). Experiments have been carried out on UK irradiated graphite waste to investigate the removal of isotopic content prior to long term storage and to assess the long term leachability of isotopes. Several leaching conditions have been developed in order to remove ^{3}H and ^{14}C from the irradiated graphite. Leach results obtained from this analysis and differences observed under varying leaching conditions are discussed. Thermal analysis of the samples pre- and post-leaching has been performed to quantify the ^{14}C and ^{3}H inventory. Finally, the suitability of the developed chemical treatments before disposal is discussed.

Keywords
Chemical treatment, thermal analysis, irradiated graphite

INTRODUCTION

The removal of key mobile isotopic species would ensure no later migration to the environment post-disposal from nuclear graphite waste. Tritium and carbon-14 are the isotopes under investigation in this research, due to a number of factors including the significant quantities of both of these isotopes present in the material provided as well as their mobility and long half lives. Tritium is a highly mobile isotope capable of radiolysis and integration into ground water systems. Carbon-14 has a long half life of 5730 ± 40 years and has the capability to be incorporated into the biosphere.

There are many possible options for chemical treatment experiments on radioactive material. The choice of the leaching conditions used is often driven by envisaged repository environment or ground water systems, sample history, sample size and availability. This gives rise to an often unique approach to the experimental conditions and length of test. Water conditions are used as a standardised leaching method with amount leached at 100 days used as a comparison. This leaching experiment focuses on ^{3}H and ^{14}C removal with minimal damage and weight loss to the irradiated graphite (or i-graphite).

This project has been incorporated within the EU Euroatom CARBOWASTE programme which focuses on 'Treatment and Disposal of Irradiated Graphite and other Carbonaceous Waste'. This programme of research has over 30 partners from all across Europe working within six work packages. The work presented here contributes to work packages 3 and 4 which include 'characterising the structure, properties and contamination of i-graphite' and 'selecting treatment options and determining decontamination factors'.

REACTOR HISTORY AND MATERIAL BACKGROUND

Two types of irradiated graphite material were used in this research. Irradiated graphite samples were taken from the British Experimental Pile Zero (BEPO) Energy Reactor, and in addition to these BEPO samples, irradiated graphite material was obtained from the Wylfa Magnox Reactor 1.

The British Experimental Pile Zero (BEPO) Energy Reactor was commissioned in 1948 and closed in 1968. BEPO was graphite moderated, air cooled and initially fuelled with natural uranium metal in aluminium cans. Two grades of enriched uranium fuel were used at a later date within the reactor. BEPO was initially used as a prototype for the Windscale piles with BEPO used predominately as a research reactor (UKAEA, Dickinson *et al.*, 1958; Thewlis, 1945).

In 1975, a four inch cylindrical sample was trepanned from the BEPO core, through the control face of the pile, steel shielding, concrete bioshield and the 20 columns of graphite blocks (Wickham, 2002). From the cylindrical graphite blocks, trepanned core samples were taken. A diamond-tipped hole cutter using no coolant or lubricant was used to retrieve these samples (Wickham, 2002). Figure 1 shows the fluence experienced by each channel (Wickham, 2002), where the sample number indicates the channel from which the graphite sample was taken. Sample 1 (Channel 1) was furthest away from the centre of the core and Sample 21 (Channel 21) was at the centre of the BEPO reactor core.

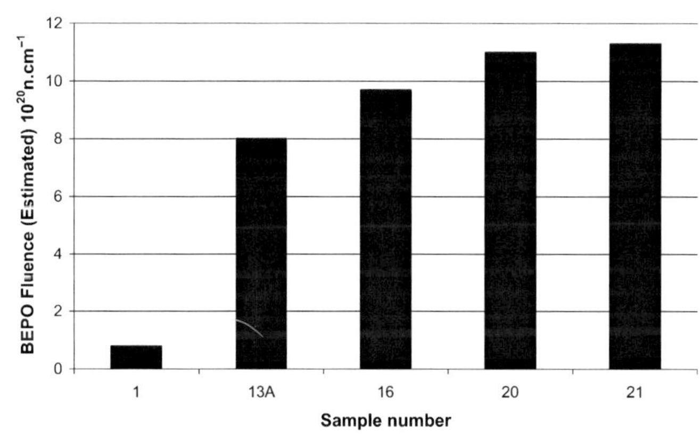

FIGURE 1: BEPO neutron fluence against channel number.

Wylfa Magnox reactors were commissioned in 1971 and consist of two reactors with a combined capacity of 980 MW. Wylfa reactors were the largest and last of the Magnox reactors to be built in the UK. Each Wylfa core has 6,156 vertical fuel channels and contains over 49,248 natural uranium Magnox-clad fuel elements; with a primary coolant of carbon dioxide. The samples issued to the University of Manchester were trepanned from the fuel channel wall of Reactor 1 channels 1413/02 and 1319/12 in April 2007. Figure 2 shows the location of these two sampling points. There are twelve bricks per column in Reactor 1 and they are numbered from the bottom of the core. Sample 1413/02 9U comes from an upper position in brick nine and sample 1319/12 8U comes from an upper position in brick eight.

Table 1 details the irradiation history of these two samples. Sample 1413/02 was used for these leaching experiments.

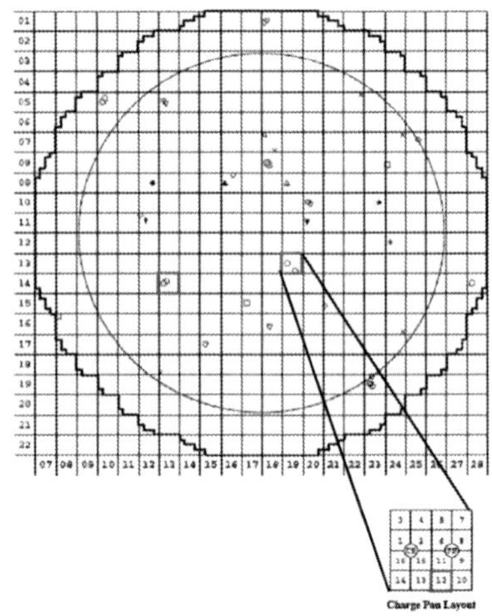

FIGURE 2: Highlighted are the samples taken in 2007 from the Wylfa reactor for this work.

Irradiation conditions	1413/02 9U	1319/12 8U
Irradiation temperature (˚C)	352	342
Adjacent fuel dose (MWD/t)	34973	37319
DIDO equivalent dose at channel wall (10^{20} n/cm^2)	34.21	37.11

TABLE 1: Irradiation conditions of Wylfa samples used in this study.

THERMAL ANALYSIS OF ^{3}H AND ^{14}C INVENTORY

Thermal analysis has been carried out on all i-graphite samples used in this study to quantify accurate inventories of ^{3}H and ^{14}C present. This analysis utilised a combustion furnace with off-gas collection using a liquid scintillation beta counting (LSC). The efficiency of the LSC was determined by sucrose standards containing ^{3}H and ^{14}C with known activities, this allowed comparison of a theoretical value to an experimentally determined LSC value. The sucrose standards containing ^{3}H and ^{14}C were also used to monitor the furnace efficiency and catalyst condition; this process is called a recovery check as a known value of activity is analysed by thermal analysis providing a comparison to the experimentally determined amount recovered. Water flushes are additionally preformed after each analysis and off-gas collection is analysed to ensure no carry over or contamination of glassware has occurred.

CHEMICAL TREATMENT RESULTS

0.5 grams of i-graphite was placed in 50 ml solution at room temperature. Each set of leachant conditions was duplicated whenever sample volumes allowed. The IAEA 'semi-dynamic' leach test approach was adopted (Wickham and Marsden, 2006), which involved 5 ml aliquots removed for analysis at each specified time point. Then 5 ml aliquots of fresh solution was then added to the leachant which ensures the overall volume remains the same throughout the test. Testing frequency was once every 7 days with isotopic release rates calculated using liquid scintillation counting. This experiment was performed for 190 days, steady state equilibrium was reached in all cases before day 90. Determination of the leach rate at the steady state would evaluate the effectiveness of pH conditions investigated on isotopic release. Samples from BEPO Channels 1, 16 and 20, and Magnox Wylfa graphite were evaluated. Leaching analysis of ^{3}H under acidic conditions is shown in Figure 3 and suggests that pH had an affect on the partial isotopic exchange of ^{3}H. Under water conditions, 2.1% of the total tritium inventory has been removed whereas under acidic conditions and pH 1 buffer solution 11.2% of the total ^{3}H inventory was leached out. One possible ^{3}H leaching mechanism proposed maybe ^{3}H isotopic exchange with the acidic species present in solution.

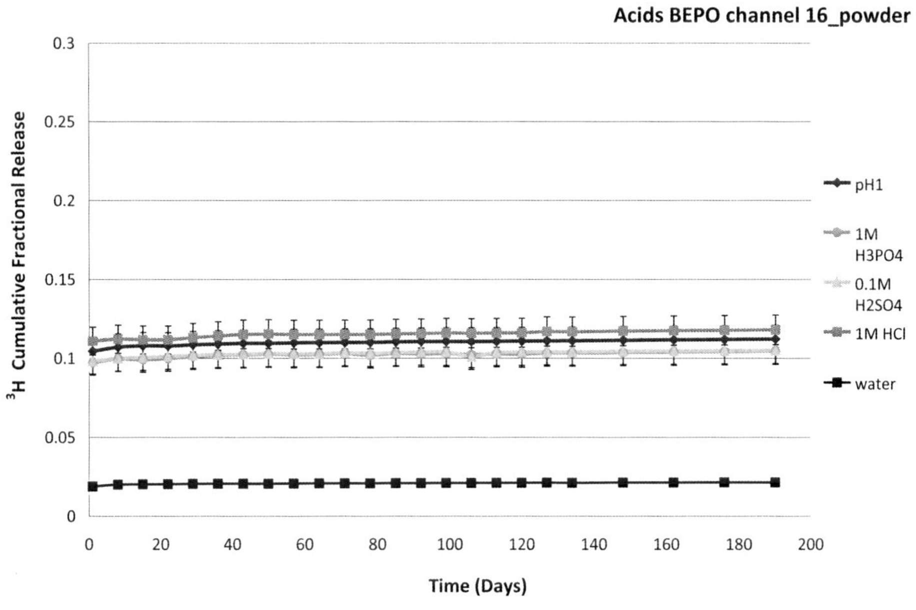

FIGURE 3: ^{3}H release under acidic conditions.

Figure 4 shows the analysis achieved from leaching powdered BEPO Channel 16 graphite under water and pH 1 conditions. It can be seen that 26% of the total ^{14}C inventory was removed under acidic conditions and 1.2% ^{14}C was leached out under water conditions. Initial results on the ^{14}C inventory present within the BEPO Channel 16 graphite supports the theory that ^{14}C may be present in both a leachable and non-leachable chemical form. A possible mechanism behind this leachable ^{14}C release could be ^{14}C-OT group exchange or $^{14}CO_2$ production.

Figure 5 shows a comparison of the analysis for BEPO Channel 16 powdered graphite under water, pH 1 and pH 13 buffers. Both water and pH 13 had overall release rate of <4% ^{3}H and pH 1 buffer released 11.2% ^{3}H. This supports the previous theory that under acidic conditions isotopic exchange of ^{3}H is the most probable mechanism the significantly reduced release rates observed under water and pH 13 conditions supports this statement.

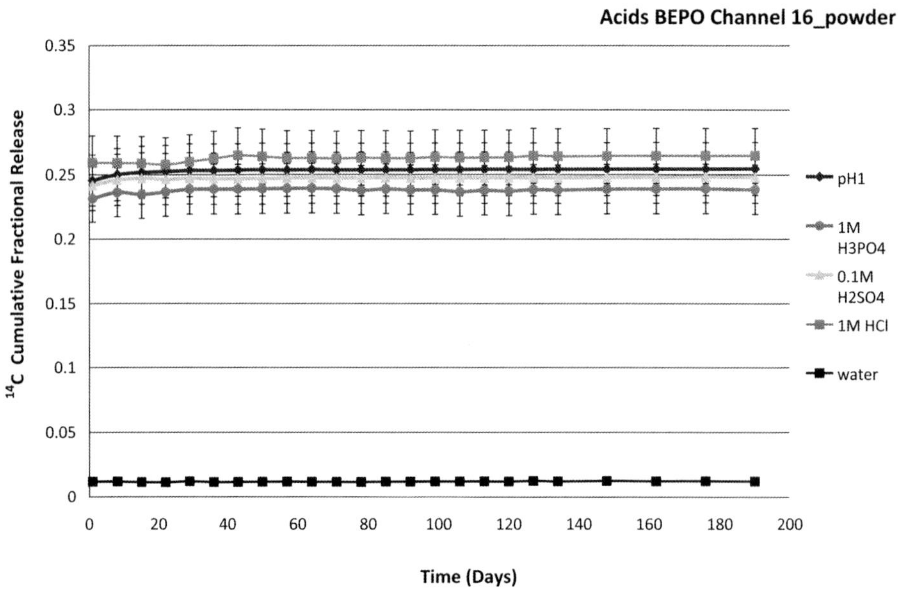

FIGURE 4: ^{14}C release under acidic conditions.

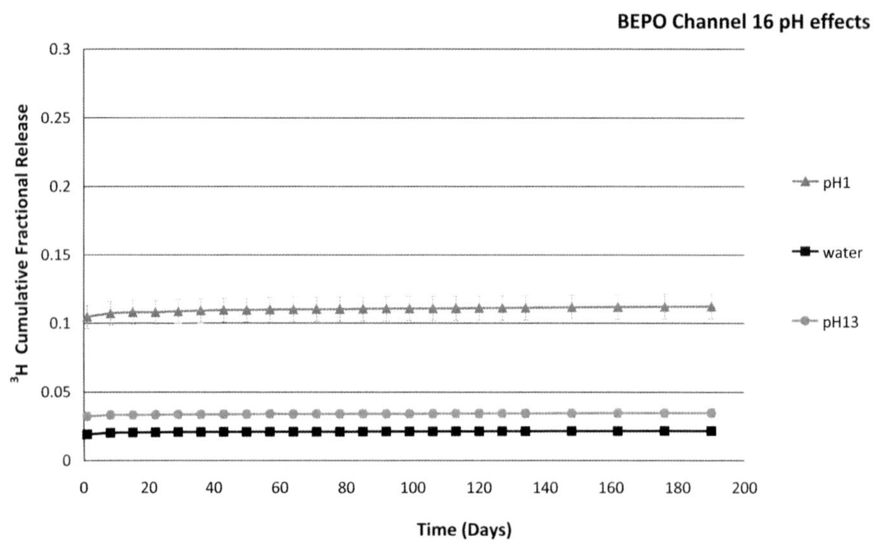

FIGURE 5: ^{3}H release under buffered conditions.

Significantly less ^{14}C was released under pH 13 buffer solution compared to pH 1 buffer solution. Figure 6 shows significant differences between these results for BEPO Channel 16 graphite with 3.5% of ^{14}C released at 190 days. A common trend observed throughout all these chemical treatment experiments is that a steady state is reached by sampling point day 90. After this time point, very little further activity is released for both ^{14}C and ^{3}H isotopes.

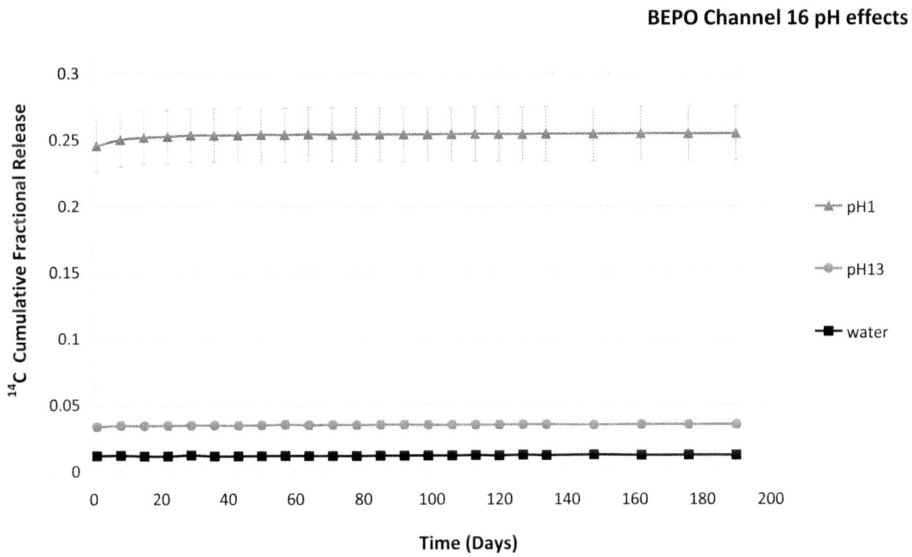

FIGURE 6: ^{14}C release under buffered conditions.

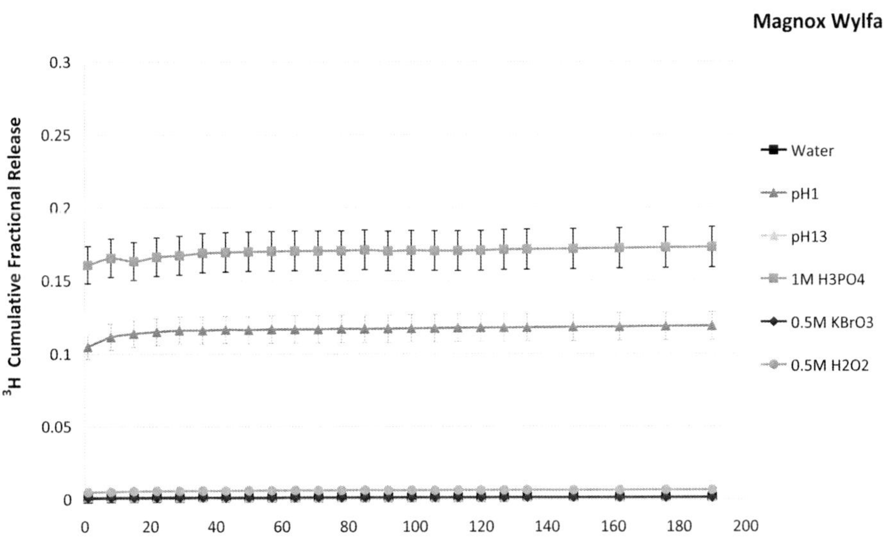

FIGURE 7: ^{3}H release under all conditions on Magnox Wylfa graphite.

Figure 7 shows the leaching results for chemical treatment of Magnox Wylfa granulated i-graphite. A preferential acidic release rate is observed with 11.9% ^{3}H released under pH 1 buffer solution and 17.3% ^{3}H released under 1M orthophosphoric acid. This result further supports the proposed mechanism of isotopic exchange for tritium release. Under water, pH 13 buffer solution, 0.5M Hydrogen peroxide and 0.5M Potassium Bromate <1% ^{3}H activity has been leached out. One driving mechanism behind this limited release rate is thought to be due to the graphite being in a granulated form not powder which could have restricted the mobility of ^{3}H. Confirmation of this mechanism is being investigated at the University of Manchester presently on other powdered Magnox graphite and determination of the surface area using BET analysis.

CONCLUSIONS

Graphite in a UK repository will make up 30% of the total proposed available (NDA, 2010). This graphite waste is currently at an ILW waste issue. This study has highlighted that under acidic chemical treatment conditions, the i-graphite tested here observed up to 17% of the total ^{3}H inventory removed. In particular, after 90 days of chemical treatment, the release rate of ^{3}H and ^{14}C is significantly reduced and in some cases <LOD of the LSC. This supports long term repository disposal conditions as mobile and leachable forms of ^{14}C and ^{3}H are quickly removed. Comparison of the results under pH 13 buffer solution shows that <4% of ^{3}H and ^{14}C may be removed. Limited release rate in conjunction with the granulated Wylfa results show that surface area may well be a rate determining step in isotope mobility. This supports the case for other engineering controls which may reduce isotopic release. These engineering controls include the thickness of the concrete barrier/container designed to reduce migration and preparation the moderator core into a suitable size for long term storage may in fact promote the release of isotopes within the structure.

Acknowledgements

The author would like to thank Dr S. Preston at Serco Assurance for collaboration and use of facilities, the team at INUTEC for training and support on using the carbolite® furnace and Dr T Wickham for guidance on the BEPO graphite. This work was carried out as part of the CARBOWASTE Program: Treatment and Disposal of Irradiated Graphite and Other Carbonaceous Waste, Grant Agreement Number: FP7-211333 and partial funding from the NDA.

References

Dickinson, J. L., Kinchin, G. H., Jackson, R. F., Lomer, W. M. & Simmons, J. H. W. (1958). BEPO Wigner energy release. *A/CONF.*

NDA (2002). Radioactive wastes in the UK: A Summary of the 2001 Inventory.

NDA (2010). Radioactive wastes in the UK: A Summary of the 2010 Inventory.

THEWLIS, J. (1945). British Experimental Pile (BEPO) - graphite supplies. In Harwell, A. E. R. E. (Ed.), National Archives (unclassified).

UKAEA Harwell project profiles: British Experimental Pile O.

von Lensa, W. (2008). CARBOWASTE: New EURATOM project on 'treatment and disposal of irradiated graphite and other carbonaceous waste'. In HTR2008 (Ed.) *Proceedings of the 4th International Topical Meeting on High-Temperature Reactor Technology.* Washington D.C., USA.

von Lensa, W., Vulpius, D. & Banford, A. (2011). Treatment and disposal of irradiated graphite and other carbonaceous waste. *ATW - International Journal for Nuclear Power.,* 56, 263.

Wickham, A. J. (2002). BEPO graphite: compilation of data relevant to test of GIRM methodology. Department of Trade and Industry.

Wickham, A. J. & Marsden, B. J. (2006). Characterisation, treatment and conditioning of radioactive graphite from decommissioning of nuclear reactors. *IAEA TECDOC-1521.*

Part C

Statistical

Statistical Modelling of the Ageing of Graphite Cores

Philip R. Maul
Quintessa Ltd., The Hub, 14 Station Road, Henley-on-Thames UK, RG9 1AY
Email: philipmaul@quintessa.org

Abstract
Statistics must be one of the most misused branches of mathematics. Nevertheless, statistical modelling techniques offer powerful tools that can be used in concert with other types of model to gain insight into the behaviour of complex systems such as the graphite cores of gas-cooled nuclear reactors. Some of the issues that need to be addressed go to the heart of the whole philosophy of mathematical models: why we need them and what they can do for us. Some key concepts for mathematical modelling in general and statistical modelling in particular are considered and applied to problems associated with the ageing of graphite cores with mathematical details only being given where these are needed to illustrate key concepts. Statistical models have a major role to play in this field, not least in helping to avoid fallacious conclusions being drawn from the use of other types of model.

Keywords
Statistical modelling, complex systems, graphite cores

INTRODUCTION

Statistics is all about what we can learn from data. Statistical modelling is all about using statistical techniques when modelling systems that include, or appear to include, an element of variability or randomness.

We begin by considering briefly some of the problems with the use of statistics in mathematical modelling and then proceed to consider issues that are central to mathematical modelling in general and statistical modelling in particular. Examples of how statistical models can help gain insight into the ageing of graphite cores are then given which show how the representation of variability is crucial for obtaining a satisfactory understanding of how the system will evolve in future.

WHY DO PEOPLE DISTRUST STATISTICS?

Statistical methods are firmly based on probability theory and many of the problems people have with statistics derive from the counter-intuitive nature of some of the results of probability theory.

Most people have come across the problem of spurious correlations. Newspapers are full of articles that purport to show a relationship between two factors, often associated with human health. If there is a statistically significant correlation between two quantities A and B, it is tempting to conclude that A causes B (or vice versa). However, the relationship may have occurred by chance, or be due to a confounding factor that affects both A and B. If we undertake 20 regression calculations for a quantity of interest against different factors that might be associated with it and undertake a statistical test at the 95% confidence level, we should not be surprised if one of the correlations appears to be significant by chance! This is an example of where statistical tests can provide useful information to support research findings, but should not be used blindly.

An example of a much more subtle effect that illustrates just how counter-intuitive statistical theory can be is given by Stein's paradox. The mathematical details are given in, for example, Young and Smith (2005) and a discussion of it has been given in a popular science journal by Efron and Morris (1997). If, for example, we know the batting averages of some cricketers playing for different cricket teams in different parts of the country and want to estimate how well they will do in their next innings, most people would assume that our best estimates would be their batting averages to date. This is indeed the case if we only have one or two batsmen, but Stein showed that when we have three or more batsmen better estimates can be made despite the fact that the performances of the batsmen are independent. This is not the place to go into details of why (and how) we can actually do better than this, but it does illustrate the general point that some results from statistical theory seem to run counter to common sense.

Another reason for distrust (particularly amongst scientists) is that statisticians often cannot even agree amongst themselves. There are many examples of such 'civil war' with the most well known being the differing views of 'frequentist' and 'Bayesian' statisticians. The basic difference is that the former believe that the 'data should speak for themselves' in any experiment whilst the latter believe that prior expectations of the outcome of the experiment are relevant. Often the differences between the two approaches do not materially affect the conclusions that can be drawn from statistical modelling; methods from both schools of thought are used in the examples of statistical modelling of reactor cores given later in this paper. When data are plentiful, statistical tests derived from the two approaches are generally similar, but when the data are sparse this is not the case. Taking a frequentist approach, it is often not possible to draw any firm conclusions from a few data points, but if the data strongly support the prior assumptions, this can be very useful if a Bayesian approach is employed. It is interesting to note that one approach to resolving Stein's paradox is obtained by taking a Bayesian perspective.

WHY USE MATHEMATICAL MODELS?

The whole point of developing a mathematical model is that it should have predictive power. We use information from historic data to develop the model, but the model is of no value unless it can be used to make predictions for other situations. Newton observed how the movement of objects was affected by imposed forces and postulated his famous laws of motion. These laws amount to a model for the movement of solid objects subject to external forces and have been shown to work very well in everyday experience. They are, nevertheless, an approximation, as shown later by Einstein. This illustrates an important point: most models (if not all) are approximations to reality, but this does not limit their usefulness in making predictions provided their range of validity is understood.

The only satisfactory way to test (validate) a model is to use it to make blind predictions. For example, if an experimentalist is given, information on the mass of a particle and its position in a gravitational field they can make predictions for its subsequent motion that can be used to test the validity of the model employed (the equations of motion). This is central to the whole scientific method.

Types of Model

Models are often considered to be of different types, but these differences are less important than sometimes suggested. Although definitions may vary, in many fields a distinction is made between models for a given system that are "empirical" and those that are "mechanistic". Empirical models are 'data driven' whilst mechanistic models are 'theory-driven'. In fact, 'empirical' and 'mechanistic' are two idealised ends of a spectrum: most models have some 'empirical' and some 'mechanistic' characteristics. Purely empirical models (*i.e.* without any physical understanding) are rare, although neural networks may be considered to be such. Similarly, no models are purely mechanistic, at least beyond the atomic scale. There is a good summary of these points from the field of pharmokinetics by workers at Manchester University in Nestorov *et al.* (1999).

Whatever the type of model, one of its most important features is the number of 'free' parameters that have to be fitted to the available data. Empirical models will tend to have more such parameters than mechanistic models, but, as argued above, all models have some empirical component to them. When a model is described, it is important that we are 'honest' about the number of such parameters.

Statistical Models

Statistical models are one sort of model that can contribute to understanding systems, particularly complex systems (and reactor cores are certainly complex), but most standard texts on statistical theory tend to concentrate on interpolation rather than extrapolation. It is much easier to determine the confidence we can have in estimating a quantity that lies in the range covered by the historic data compared with estimating something that may happen in the future, outside this range. A simple example of this is given in Figures 1 and 2. Here, there are just five data points for the values of a quantity of interest for years 1-5. Figure 1 shows a simple linear fit to the data (a model with three fitted parameters: the intercept and slope of the line and the variance of the error term); this seems to capture the generally increasing trend with time, and it is reasonable to expect that when extrapolated forward the model would give a sensible prediction for the values to be expected in years 6 and 7. Figure 2 gives a polynomial fit to the data; this has five free parameters (one for each data point) and apparently provides a much better fit to the historic data, but when extrapolated outside the range of the historic data we can have little confidence that the predictions will be of any use. Given the quality and quantity of the historic data, the simpler model is more useful and is likely to have greater predictive power.

Functional Forms

Starting with a 'blank sheet of paper' there is effectively an infinite number of possible functional forms that could be considered for statistical models of a complex system. However, for these models to be useful they need to be linked in some way to what is known about the key processes that are operating.

As an example, consider the basic form of the models that has been used to represent the bore cracking of graphite bricks (Maul *et al.*, 2011). The risk of a brick undergoing a first bore crack is taken to be given by:

$$h = u\,g(\tau) \quad \tau > \tau_0 \qquad [1]$$

Here τ is a measure of the brick 'age', $h\delta\tau$ is the probability that the brick will crack in a small interval $\delta\tau$ after an initial period τ_0, and $g(\tau)$ is a low order polynomial in τ. At first sight, it might appear that the chosen form does not reflect much physical understanding, but:

- It is known with a fair degree of certainty that there was a period of reactor operation before any bore cracks were seen: hence the inclusion of the parameter τ_0.

- A suitable age measure to use is suggested to be related to the radiation damage dose incurred by the brick.

- There is an extended period over which the cracking risk is significant (determined by the form of $g(\tau)$); this reflects the variability in graphite properties which means that all bricks of the same 'age' will not crack at the same time.

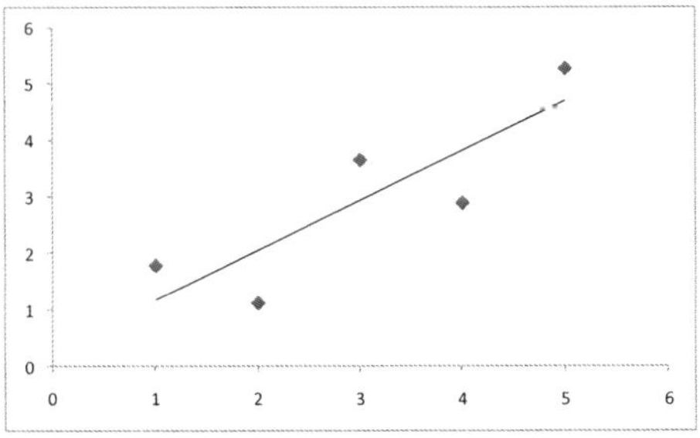

FIGURE 1: A linear regression fit.

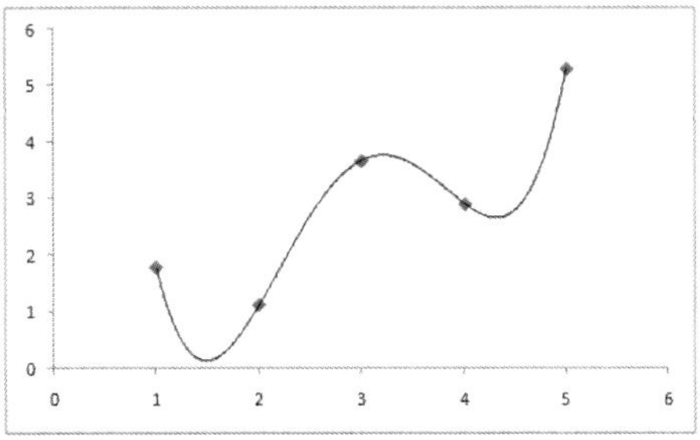

FIGURE 2: A polynomial regression fit.

Model Complexity

As shown in the regression fit example above, the most complex model is not necessarily the best. In general, there is a trade-off between improving the fit to the historic data and increased complexity. There are a number of information criteria in statistics that aim to quantify this trade-off. They are of the general form:

$$I = 2(l - p), \qquad [2]$$

where I is the information criterion, l is the maximum likelihood function, which reflects how well the model fits the data, and p is a penalty function that increases with model complexity. Sometimes the information criterion is defined as the negative of this expression. The best known criterion is that due to Akaike, the AIC, where the penalty function is given by the number of free parameters. An idea that is related to another information criterion (Takeuchi's model-robust information criterion, TIC) is cross-validation. The model is fitted with one data point left out each time and an average taken of the n log-likelihoods. The model with the highest average log-likelihood is preferred.

There are a number of technical issues associated with the use of these criteria which derive from the fact that we do not actually know what the 'true' model is - life would be much easier if we did - but they represent a practical way of evaluating what the appropriate level of model complexity is for the quality and quantity of data that are available.

Using these criteria, it may be possible to reject some candidate models which show a much poorer fit to the data than the best performing models. Statistical tests are availale to support this process, and the whole area of model selection and the use of information criteria is discussed in more detail in Claeskens and Hjort (2008).

An illustration of how the appropriate level of model complexity can increase as more data become available is shown in Table 1. For the statistical models for axial bore cracks at the Hartlepool and Heysham 1 power stations very few data were available in 2005, and the best model was a very simple one with just two free parametrers. Now that much more observational data are available more complex models fit the data best.

Reactor	Inspection Date	Burnup (TWd)	Fitted parameters for best model	New Single Cracks	New Double cracks	Position of Outcome in Prediction CCDF (%)
HYA R2	May 05	7.32	2	8	0	8
HYA R1	Sep 05	7.82	3	3	1	66
HRA R1	Jan 06	7.41	2	4	0	100
HRA R2	Jun 06	7.97	2	2	0	60
HYA R1	Dec 06	8.34	2	2	0	34
HYA R1	Jun 07	8.55	2	2	0	57
HRA R1	Sep 07	8.05	3	3	1	77
HYA R2	May 08	8.29	4	2	1	48
HYA R2	Jun 08	8.29	4	0	1	19
HYA R1	May 09	8.77	5	6	0	46
HRA R2	Oct 09	8.73	5	5	3	9
HYA R1	Jun 10	9.26	5	2	2	79
HRA R1	Jan 11	9.00	5	5	0	63
HRA R1	Jun 11	9.17	5	6	3	22

TABLE 1: Inspections at Heysham 1 (HYA) and Hartlepool (HRA): full-length axial cracks.

Predictive Power

When a model is used to make a blind prediction of an experimental observation, it is necessary to define what outcomes would be 'expected' and what would be 'surprising'; a single 'best estimate' prediction is of little use as there is then no way of evaluating the significance of any difference between the prediction and the measurement. If the prediction is in the form of a probability density function (PDF), then the position of the outcome in the cumulative density function (CDF) can be determined; it will lie somewhere in the range 0 to 1. For a single continuous quantity we might consider, for example, that if the observation lies in the central 95% of the prediction PDF (*i.e.*, it lies between 0.025 and 0.975 in the CDF) then this is 'expected', but it is 'surprising' if it lies outside this range. If however, we undertake a number of such predictions, we would expect over time that some of the measurements would in fact lie in the 'unexpected' part of the range. In the long run, if we have a 'good' model, the values of the outcomes in the CDFs should be compatible with a uniform distribution over the range 0 to 1: there are statistical tests that can be employed to see if this is the case. It is important to ensure that possible measurement errors have been taken into account when making these comparisons.

When more than one quantity is to be predicted and these are discrete, the analysis is slightly different. For example, in Table 1 the numbers of single and double axial cracks are predicted, and the position of the measurement in the prediction ranges is expressed in terms of the complementary cumulative density function (CCDF); here the percentage of outcomes that are at least as likely as the outcome in question is given, so that the predicted most likely outcome would have a value of 100%, and very unlikely outcomes would have values close to 0%. The technical issues involved in such comparisons are discussed in Maul *et al.* (2011).

When making predictions for graphite properties, it is important to consider that some properties may be correlated. Figure 3 gives an example of comparisons between the predictions made using statistical models for the 2008 Hinkley Point B Reactor 4 (HPB R4) trepanning campaign. The residuals (the difference between the model central predictions and the measurements) are shown for dynamic Young's modulus (DYM) and flexural strength (FS). The contours indicate zones outside of which the given percentage of results is expected. Here the observed residual correlation coefficient was 0.92, very similar to the predicted value of 0.90.

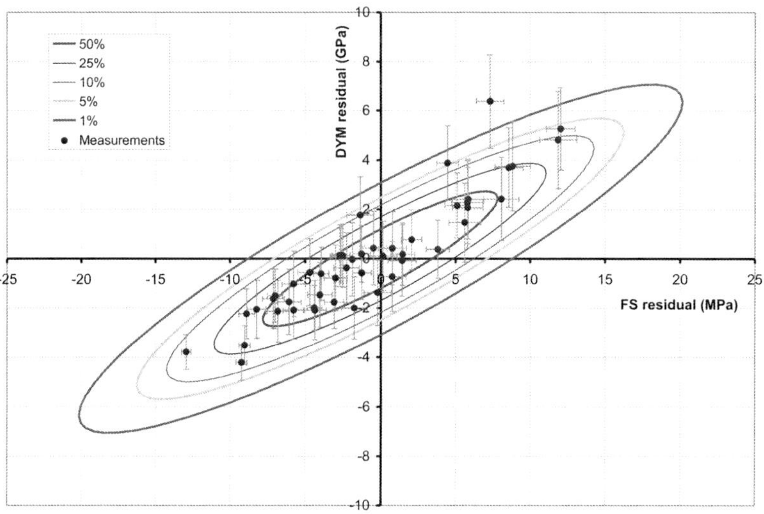

FIGURE 3: Correlation plot of DYM residuals against FS residuals for the 2008 Trepanning Campaign at Hinkley Point B Reactor 4.

There are various methods for quantifying the predictive power of different models, but it is important to reflect the aspects of the model predictions that really matter when using these. We may, for example, not be too concerned if a quantity of interest is over-predicted, but may wish to ensure that it is not under-predicted. This perspective is expressed mathematically in terms of a loss function, which represents the perceived cost associated with a prediction being 'wrong' (which gets bigger as the prediction gets increasingly 'wrong'). The concept of loss functions is discussed in, for example, Young and Smith (2005).

Uncertainty and Variability

In complex systems in general, and graphite cores in particular, it is essential to make a clear distinction between variability and uncertainty. Variability is an irreducible characteristic of the system whilst uncertainty is a consequence of a lack of knowledge about the system which may be reduced by gathering more data or improving understanding in other ways.

Most analyses of uncertainty consider a single model and concentrate on the uncertainties associated with estimating the values of the model parameters. It is important to acknowledge that this is not the only, or necessarily the dominant, source of uncertainty. Wherever possible multiple models that are consistent with the available data should be considered in order that conceptual model uncertainty can be investigated. In modelling the ageing of reactor cores, this can often be the most important source of uncertainty.

Historically, variability and uncertainty have not always been distinguished satisfactorily in models for the ageing of graphite core components. There is actually variability at different scales and experience with statistical models has shown the importance of separating within-brick variability from between-brick variability; simple models that do this consistently out-perform more complex models that do not. The simplest version of these models considered is the following form:

$$y_{ij} = \Phi + u_j + \varepsilon_{ij} \quad [3]$$

where the index i refers to the relevant brick and j to the sample within that brick. Φ is the model for the central trend behavior, with the parameters u representing brick-level variability and ε sample-level variability. More complex forms can be considered, for example where the two terms representing variability are weighted by dose or other combinations of variables. An indication of the relative importance of within-brick and between-brick variability can be obtained by considering the variance of these terms.

For bricks that have been trepanned in the past, information is available which can be used to provide estimates of the brick-specific parameters u. For bricks where this is not the case, no such information is available and so the uncertainty in making predictions is greater. Figure 4 shows an example of the use of these models to predict measured values of DYM in a materials test reactor experiment. Where previous trepanning information is available for the sample in question, the prediction intervals are smaller than where no such information is available.

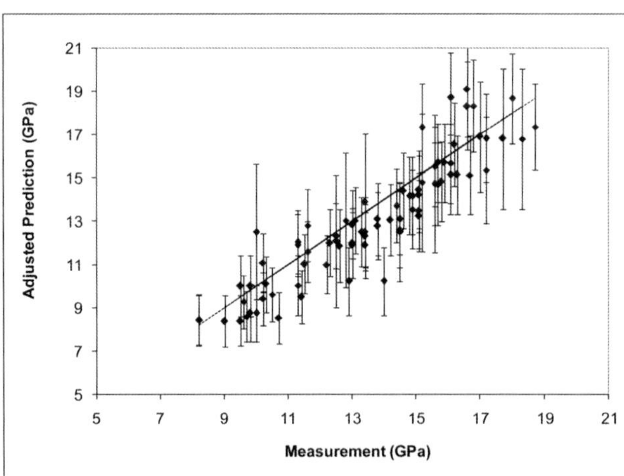

FIGURE 4: Comparisons between predictions for DYM values and measurements for ex-reactor samples from HPB in Blackstone Phase 1.

FORECASTING CORE BEHAVIOUR

Maximum likelihood estimates can be used to provide a single set of model parameters for forward projections of core behaviour. In addition, probability density functions of the parameter values can be estimated by using Bayesian updating. This process requires the specification of a prior distribution for the parameters, reflecting what is known before the data are considered. The aim will generally be to use a non-informative prior, representing a lack of knowledge before the data are considered. The choice of such a prior is not as straightforward as might be expected, but once sufficient data are available, model forecasts will not generally change very much with the prior that is employed. An exception can be when information on the tails of the PDF of the property of interest is considered; because few data will lie in the region of interest, the assumptions made in the prior can still dominate the forecast. This is an example of a general problem; extreme caution should be exercised when trying to draw conclusions far into the tails of the distributions for quantities of interest.

Because of the large variability in the properties of graphite bricks, uncertainties in the forecasts arising from uncertainties in the estimation of the parameter values of a particular model are often dominated by the effects of variability. However, this can lead to overall uncertainties being underestimated. Using different types of model to forecast future core behaviour helps to address conceptual model uncertainty and so give a better representation of the true uncertainties than can be obtained from the use of a single model. This helps to define the period over which forecasts can be relied upon; one can have greater confidence when forecasts produced by different models are compatible.

Acknowledgements
The author would like to acknowledge the many contributions from Peter Robinson of Quintessa Ltd and Paul Northrop of University College London to the ideas presented in this paper. The agreement of EDF Energy to the use of the illustrative calculations for its Advanced Gas Cooled Reactors is gratefully acknowledged.

References

Claeskens, G. and Hjort, N. L. (2008). Model selection and model averaging. Cambridge University Press. ISBN 978-0-521-85225-8.

Davison, A. C. (2003). Statistical models. Cambridge University Press. ISBN 0-521-77339-3.

Efron, B., and Morris, C. (1977). Stein's paradox in statistics. *Scientific American*, 236 [May], 119-127.

Maul, P. R., Robinson, P. C. and Northrop, P. J. (2011). Statistical modelling of graphite brick cracking in advanced gas-cooled reactors. *Appl. Statist*, 60, [3], 339-353.

Nestorov, I., Rowland, M., Hadjitodorov, S. T. and Petrov, I. (1999). Empirical versus mechanistic modelling: comparison of an artificial neural network to a mechanistically based model for quantitative structure pharmacokinetic relationships of a homologous series of barbiturates. *AAPS Pharmsci.*, 1, [4], 1-9.

Young, G. A. and Smith, R. L. (2005). Essentials of statistical inference. Cambridge University Press. ISBN 0-521-83971-8.

Probability Modelling of Density Variability of Radiolytically Oxidised Graphite

Robert Moskovic
Magnox LTD, Oldbury Technical Centre, Oldbury Naite, Thornbury, South Gloucestershire BS35 1RQ, UK.
Email: robert.moskovic@magnoxnorthsites.com

Abstract
A Magnox reactor core is constructed of twelve to thirteen layers of Pile Grade A graphite bricks which serve as a moderator. There are fuel channels within the centre of each brick and interstitial channels at the corners of some sets of four bricks. The microstructure of a brick is heterogeneous comprising a mixture of filler particles, matrix and porosity. The density of virgin graphite varies both between and within bricks. Carbon dioxide is used as coolant and gives rise to radiolytic oxidation causing loss of mass and the associated decrease of density. The changes of density are measured on core samples trepanned throughout the life of the reactor. The statistical structure of density data is hierarchical and needs to be modelled using a probability model in which both systematic and random components of density variability are quantified. The paper illustrates how this problem can be quantified using data obtained from Magnox gas cooled reactors.

Keywords
Radiolytically oxidised graphite, hierarchical density data, mixed effects models.

INTRODUCTION

All power generating nuclear reactors in the UK, except one, are gas cooled by carbon dioxide. Two generations of gas cooled reactors have been built: Magnox, initially, followed by Advanced Gas-cooled, later. The cores of the older Magnox reactors are constructed of Pile Grade A graphite (PGA) bricks which function as a moderator. The bricks were manufactured from a mixture of dry calcinated coke and liquid pitch binder. The raw materials were extruded, baked at a temperature of approximately 800°C to remove volatile substances, pitch impregnated and finally graphitised at a temperature of approximately 2800°C (Nuclear Graphite Production, 1957; Kelly, 1981). The microstructure of graphite, produced by these methods consists of relatively large graphitised filler particles (*i.e.* 100 to 1000 μm scale), dispersed in a matrix of finer (below ~10μm) graphite crystallites from "flour" (fine calcined filler particles) and graphitised pitch binder. The porosity arising from gas evolution and accommodation cracking due to thermal strains results from both baking and graphitisation (Nuclear Graphite Production, 1957). This generated a heterogeneous microstructure that is a mixture of filler particles, matrix and porosity. During service, the graphite looses mass due to radiolytic oxidation by CO_2. The changes of density caused by radiolytic oxidation are measured by trepanning cores, approximately 12 mm in diameter and 20 mm long, from graphite bricks after different periods of service. The trepanned cores are subdivided into three slices which are used to measure density. Both the heterogeneous microstructure and change of mass due to oxidation cause variability of density. The density of virgin graphite varies both within and between bricks. The variability of density is due to systematic, fixed, and random effects. The contribution of different effects to variability of graphite can be quantified by interrogating the data by statistical analysis.

(a) (b)

FIGURE 1: Shows (a) a trepanned cylinder and (b) slice obtained after sectioning.

The statistical analyses of density data traditionally model the potential effects on trends in the density variability caused by covariates: adjacent fuel dose, core height, channel type and slice number, as systematic (fixed) effects and random variability, *i.e.* error, represented by residual error estimated as a difference between the measured and the predicted values of density. This approach is extended in this paper by quantifying the random variability in the fitted parameters to which cannot be accounted for by systematic effects.

DATABASE

The samples for density measurements were obtained from trepanned core cylinders with their axis perpendicular to the core height. The cylinders, Figure 1(a), were subdivided into three slices each of which was 6 mm high and 12 mm in diameter. The slices were numbered from 1 to 3. The increase in the slice number indicates increasing distance from the channel wall. Figure 1(b) illustrates one of the slices. The data can be subdivided to smaller subsets according to reactor number, channel type and slice number from channel. The material nearest to the wall of the fuel channel receives a higher dose than the material that is further away. As result, the behaviour of Slice 1 data may be different from the other two slices from the same cylinder.

STATISTICAL MODELLING

Density reduction occurs due to radiolytic oxidation. The physical mechanisms of radiolytic oxidation are well understood (Best, 1985). The process is dependent on both total energy deposited and the mass of CO_2 in a pore which in turn depends inversely on local graphite temperature. The total energy deposited is principally due to gamma and fast neutron dose irradiation. Both are assumed to correlate with the adjacent fuel dose which is measured in terms of burn up in MWd/t. Within a brick layer, both the temperature and dose vary by a small amount. Their variation is greater between the core layers. Hence a relationship for the dependence of graphite density on temperature was modelled as the effect of reactor core height on the slope of density dependence on the adjacent fuel dose.

Fitting a statistical model to data involves a trade off between the goodness of fit and extent of uncertainty in the estimated parameters. The goodness of fit increases with the complexity of the model as represented by increasing number of variables and the associated parameters in the model. Inclusion of unnecessary variables in the model increases the uncertainty in the estimated parameters. The goodness of fit of the model is indicated by R^2 - a ratio between the variation explained by the model and the total variation in the data – and the maximum likelihood. The R^2 can be used when the random scatter of the data is modelled by a normal distribution with a constant variance. The least possible number of variables required to give a good explanation of the variability of the data can be estimated by generating candidate variables and identifying statistically significant variables. This has been effected initially by stepwise regression used to select the most significant variables from a large number of candidate variables. Performing the regression and testing the significance of fitted parameters involves the assumption of normal distribution and random errors with a zero mean and a uniform variance. These assumptions were tested by performing diagnostic tests involving plotting standardised residuals against the predicted values of density from regression and explanatory variables. A further testing of normality was performed by plotting quantiles expected from a standard normal distribution against quantiles for the residuals obtained from regression analysis.

Linear Mixed Effects Models

In the regression analyses performed by stepwise regression, random variability is captured by random residual errors represented by ε_i as illustrated in Equation [1]. The variability in density arising between the reactor graphite core layers is modeled by using terms for the adjacent fuel dose and its product with core height as explanatory variables. The parameters estimated for these explanatory variables are constants which are regarded as fixed effects. An alternative approach is that some of the coefficients for the explanatory variables, in the regression model, are viewed as realisations from a probability distribution with a zero mean and, for example, standard deviation of σ_1 giving $b_{1j} \sim N(0,\sigma_1^2)$. If the dependence of density on explanatory variables in a fixed effects model is given by:

$$\rho_i = \beta_0 + \beta_1 d_i + \beta_2 h_i d_i + \sigma\varepsilon_i \qquad [1]$$

where ε_i is a random error from a normal distribution: $\varepsilon_i \sim N(0,\sigma^2)$, then a similar mixed effects model, when all three coefficients exhibit random variability, is given by:

$$\rho_i = \beta_0 + b_{0j} + \left(\beta_1 + b_{1j}\right)d_i + \left(\beta_2 + b_{2j}\right)h_i d_i + \sigma\varepsilon_i \qquad [2]$$

where j indexes the subject which is the term for an explanatory variable giving rise to random variability of the regression coefficients. In the context of density measured on trepanned samples from a reactor core, reactor number, layer, channel and brick can all be subjects in the regression model. The terms b_{0j}, b_{1j} and b_{2j} are random values from normal distributions with mean of zero and standard deviations σ_0, σ_1 and σ_2, respectively. For example, $b_1 \sim N(0,\sigma_1^2)$. Hence, the random probability distribution for each regression coefficient is obtained by estimating the standard deviation. For the model given by Equation [2], σ_0, σ_1 and σ_2 are between and σ is within the subject standard deviation. The mixed effects models generate standard deviations both within and between the subjects.

Combination of these two components of standard deviation gives the total standard deviation which for the model in Equation [3] is given by:

$$\sigma_T = \left(\sigma_W^2 + \sigma_B^2\right)^{0.5} \qquad [3]$$

where σ_W^2 is the variance within the subject and is given by σ^2 in Equation [2] and σ_B^2 is the variance between the subjects and is given, for the model in Equation [2], by:

$$\sigma_B = \left(Var(b_0) + Var(b_1)d^2 + Var(b_2)(hd)^2 + 2Cov(b_0b_1)d + 2Cov(b_0b_2)hd + 2Cov(b_1b_2)hd^2\right)^{0.5}$$

[4]

Equation [4] shows that σ_B depends on the variance, *Var*, of and the covariance, *Cov*, between the coefficients for the random effects. The fixed effects parameters, βs, in Equations [1] to [3], are subject to uncertainties known as their standard errors.

RESULTS

The data for the interstitial and fuel channels were analysed separately. The initial assessment of the data by stepwise regression of six different datasets associated with different slice number and reactor combination showed that the data for fuel channels were well described by a model based on:

$$\rho_i = A_0 + A_1 d_i^2 + A_2 h_i d_i^2 + A_3 h_i^2 d_i^2 + \sigma\varepsilon_i \qquad [5]$$

And for interstitial channels:

$$\rho_i = A_0 + A_1 d_i + A_2 h_i d_i + \sigma\varepsilon_i \qquad [6]$$

Where ρ is the density, A_1, A_2 and A_3 are parameters fitted by regression, d_i is the adjacent fuel dose, h_i is the core height, i indexes the data, ε_i are standard deviates from a normal distribution with a zero mean and variance of 1 and σ is the standard deviation estimated from regression. All explanatory variables in Equations [5] and [6] have a partial R^2 value greater than 0.01 probability and therefore the fitted parameters for these equations are statistically significant. The partial R^2 values give a contribution to the goodness of fit specific to a single explanatory variable. Examination of the residuals showed some departure from the assumptions made in the statistical analysis. Plots of the residuals showed that the magnitude of the scatter varies between the core layers and that it also increases with both decreasing fitted value of density and increasing adjacent fuel dose. Plots of the residuals against the reactor core layer and the channel showed variation of the mean values for the fuel channel datasets and Reactor 2 interstitial channel data set.

To improve the fitted regression models, further statistical analysis were carried out to obtain a more complex relationship for variance and to fit mixed effects models in which there were two components of variance. The standard deviation obtained from this analysis had the form:

$$s_w = s_{wl} d_i^m \qquad [7]$$

where s_w is the standard deviation within the brick which is a product of two weights: s_{wl} and d_i^m dependent on the reactor core layer and the adjacent fuel dose, respectively.

Dependence of the regression coefficients, in Equations [5] and [6], on random effects due to reactor core layer, channel and slice number was investigated by fitting different statistical models and then testing their significance. The models were fitted by using the principle of both restricted and maximum likelihood. The statistical analyses showed that:

- Interstitial channel data for all three slices could be pooled and analysed as a single reactor specific set.
- For Reactor 1, interstitial channel data, random effects due to reactor core layer and channel are not statistically discernable. The slice variability of the intercept is significant and is modelled by random effects.
- For Reactor 2, interstitial channel data, random effects due to reactor core layer on the intercept, A_0 in Equation [5], and channel on the parameter for the adjacent fuel dose, A_1 in Equation [5], are statistically discernable and are modelled by random effects.
- Fuel channel data for both reactors were analysed separately as slice specific datasets. The intercept, A_0 in Equation [6], varies between reactor core layers and the regression parameter for the adjacent fuel dose squared, A_1 in Equation [6], varies between the channels. These parameters were modelled by random effects.

The derived equations obtained for Reactor 2 for all three slices of interstitial channel data and slices 1 and 2 for fuel channel data are illustrated below.

$$\rho_i = \left(1.73 + 0.0119\varepsilon_{lj}\right) - \left(0.113 + 0.00943\varepsilon_{cj}\right)\times 10^{-4} d_i + 0.604\times 10^{-6} h_i d_i + s_{wl}\left(d_i \times 10^{-4}\right)^{0.893} + s_{wl}\left(d \times 10^{-4}\right)^{0.893} \varepsilon_i \qquad \text{interstitial channels [8]}$$

Where the standard deviations for the random distributions of the parameters are given in blue, ε_{lj} and ε_{cj} are standard normal from normal distributions with a zero mean and standard deviation of 1.

$$\rho_i = \left(1.73 + 0.0137\varepsilon_{lj}\right) - \left(0.0268 + 0.00154\varepsilon_{cj}\right)\times 10^{-8} d_i^2 + 0.006\times 10^{-8} h_i d_i^2 - 0.00052\times 10^{-8} h_i^2 d_i^2 + s_{wl}\left(d \times 10^{-4}\right)^{-0.125} \varepsilon_i \qquad \text{fuel channels slice 1 [9]}$$

$$\rho_i = \left(1.67 + 0.0106\varepsilon_{lj}\right) - \left(0.0265 + 0.00136\varepsilon_{cj}\right)\times 10^{-8} d_i^2 + 0.00477\times 10^{-8} h_i d_i^2 - 0.00040\times 10^{-8} h_i^2 d_i^2 + s_{wl}\left(d \times 10^{-4}\right)^{-0.323} \varepsilon_i \qquad \text{fuel channels slice 2 [10]}$$

The values of reactor core layer dependent weights for within brick variance are given below.

Layer	2	3	4	5	6	7	8	9	10	11
R2 IC slices 1-3	0.015	0.008	0.011	0.011	0.008	0.010	0.010	0.021	0.027	0.013
R2 FC slice 1	0.037	0.030	0.037	0.042	0.034	0.042	0.040	0.031	0.043	0.033
R2 FC slice 2	0.031	0.022	0.042	0.054	0.047	0.071	0.052	0.035	0.039	0.034

TABLE 1: Weights for within brick standard deviation, s_{wl}, for different reactor core layers.

Equations [8], [9] and [10] can be used to calculate the global mean predicted density from the parameter and variables for fixed effects. Calculation of global standard deviations can be carried from standard deviations for random effects in these equations and within brick standard deviations using layer dependent weight, in Table 1, and the adjacent dose dependent weight in the equations. Calculation of global standard deviation is illustrated below for Reactor 1 fuel channels slice 1, core layer 4.

$$s = \left(0.0137^2 + \left(0.00154d^2 \times 10^{-8}\right)^2 + \left(0.037\left(d \times 10^{-4}\right)^{-0.125}\right)^2 \right)^{0.5} \qquad [11]$$

Measured and predicted graphite densities are compared in Figure 2 for Reactor 2 fuel and interstitial channels.

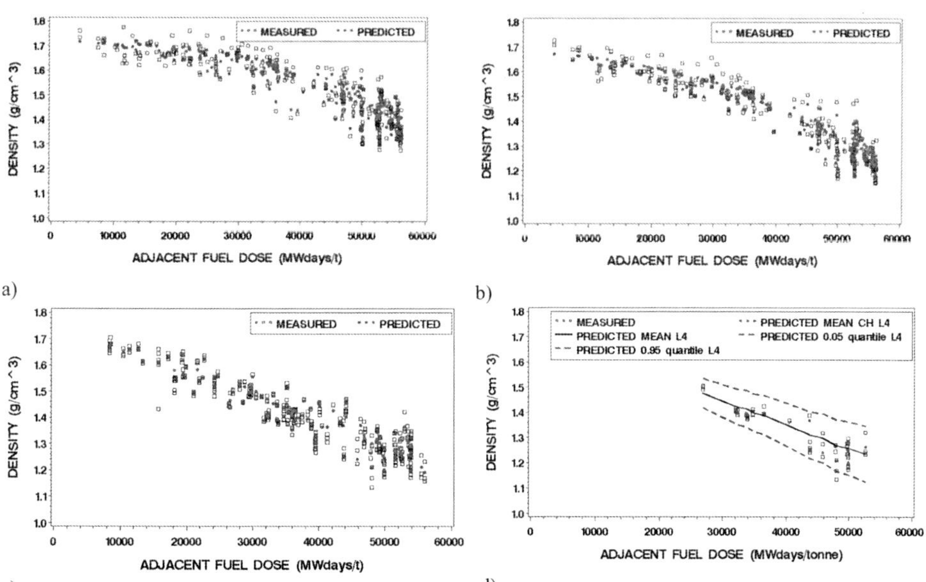

FIGURE 2: Comparisons of measured and predicted densities for Reactor 2: (a) fuel channel slice 1; (b) fuel channel slice 2; (c) interstitial channel slices 1-3; and, (d) layer 4 specific predictions for interstitial channel slices 1-3.

Diagnostic plot of normalised residuals for Reactor 2 fuel channel slice 2 data are illustrated in Figure 3. This shows that the residuals are uniformly scattered and evenly distributed around zero when plotted against the adjacent fuel dose and reactor core layer. The patterns of the normalised residuals shown in Figure 3 are consistent with the assumptions made in the statistical analysis that the errors are uniformly distributed and have a zero mean.

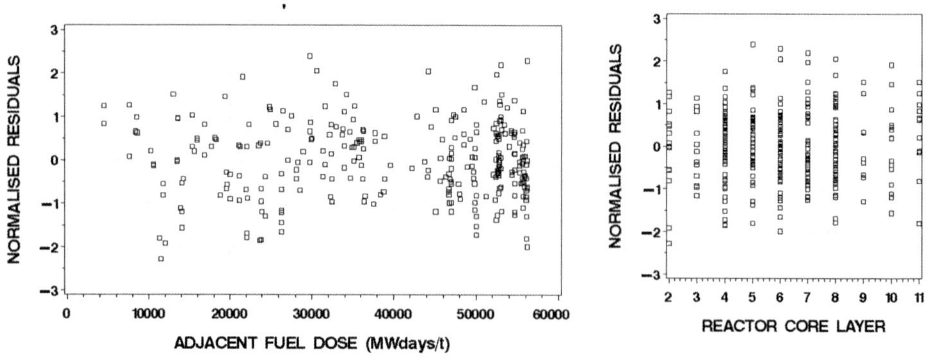

(a) (b)

FIGURE 3: Plots of normalised residuals for Reactor 2 fuel channel slice 2 data against (a) the adjacent fuel dose; and, (b) reactor core layer.

DISCUSSION

Graphite density data measured on trepanned samples from reactor core have been statistically analysed. The data have a hierarchical structure and therefore have been analysed using mixed effects models. Use of these models has some potential advantages which are outlined below. The data obtained on samples from the same channel, core layer or brick have more in common than data obtained on samples from different channels, layers or bricks. In statistical terms, such data are correlated. Mixed effects models take account of this correlation. Factors influencing the choice of the model depend partly on the structure of the data. The effect of a particular channel, core layer or brick on density can be modelled either by a fixed or a random effects model. The interpretation to be put on the results of the statistical analyses is different in these two cases. When fixed effects are fitted then interference can only be formally applied to layers, channels or bricks for which measurements were made.

If random effects are fitted then an estimate of the standard deviation is obtained for the probability distribution of the mean coefficients and inference can be applied confidently to the whole population. The results of mixed effects analysis are more appropriate to hierarchically structured data. A further difference is that in a random effects model estimates of the mean coefficients for different subjects are closer to the mean than their fixed effects counterparts. This helps to avoid the potential problem of extreme parameter estimates occurring due to chance when estimates are based on small data sets. Estimates based on small numbers will sometimes appear as outliers compared with other data due to chance variation. However, it should be noted that assumptions are made in fitting statistical models. Hence conclusions made from statistical analysis are dependent on the assumptions that have been made being valid. This can be established by diagnostic testing.

CONCLUSIONS

1 Statistical analysis of graphite density data measured on trepanned samples from reactor core has been performed using the principle of least squares to identify statistically significant variables. Mixed effects models have been fitted using the principle of restricted likelihood.

2 Data for the interstitial channels have been pooled into a single set and analysed together. Data for the fuel channels have been analysed as slice specific sets.

3 Statistical analyses showed that the systematic, fixed, effects on density are given by a relationship for density dependence on the adjacent fuel dose and its cross-product with reactor core height for the interstitial channels and by its dependence on the adjacent fuel dose squared and its cross products with reactor core height and that squared.

4 Estimates of standard deviation were obtained to quantify the random variability of some of the parameters by a normal probability distribution. The intercept for Reactor 1 interstitial channel data varies with slice number. The intercept for Reactor 2 interstitial channel data and for both reactors of fuel channel data varies with core layer. For the same datasets, the parameter for the adjacent fuel dose or that squared varies with channels.

5 The standard deviation for the random scatter within the brick depends on two weights which account for random variation due to reactor core layer and the adjacent fuel dose.

Acknowledgements
The permission of Magnox Limited to publish this paper is gratefully acknowledged.

References

Best, J. V., Stephen, W. J. and Wickham, A. J. (1985). Radiolytic graphite oxidation. *Progress in Nuclear Energy*, 16, [2], 127-178.

Hodgkins, A., Marrow, T. J., Wootton, M. R., Moskovic, R. and Flewitt, P. E. J., (2010). Fracture behaviour of radiolytically oxidized reactor core graphite: a view, *Mater. Sci. Technol.*, 401, 899-907.

Kelly, B. T. (1981). Physics of Graphite. Applied Science Publishers Ltd, Barking, UK. ISBN 0-85334-960-6.

Nuclear Graphite Production, *Nucl. Eng.*, (1957), 175-178.

Statistical Analysis of Bore Cracking in AGRs

Emma Tan, Kevin McNally and Nick Warren
Mathematical Sciences Unit, Health and Safety Laboratory, Buxton, UK. SK17 9JN
Email: emma.tan@hsl.gov.uk

Abstract
The structural integrity of Advanced Gas-cooled Reactors (AGRs) is important for their continued long-term safe operation. A critical part of the safety case is the inspection of graphite cores. Statistical methodologies for quantifying the number of cracked bricks have been developed by the Health and Safety Laboratory (HSL). Previously, simplistic statistical models of cracking were applied. A key assumption of earlier models was that the risk of cracking was the same across all bricks and channels. However, the accumulation of more recent inspection data has allowed a fuller investigation. Statistical models that allow the risk of cracking to vary between layers and channel locations have been developed. In addition, it is anticipated that the rate of bore cracking will decrease or even that it will cease in some regions. Therefore, probability distributions that allow the risk of cracking to eventually decline have been investigated. Statistical models have been fitted within a Bayesian framework using Markov Chain Monte Carlo methods. This paper describes various frameworks for modelling the relative risks of cracking. In addition, forecasts of the number of cracked bricks in the short-term future at Heysham A (HYA) and Hartlepool (HRA) are presented.

Keywords
Modelling, cracking, irradiation

INTRODUCTION

TV inspections of graphite cores have been carried out in AGRs as an important part of the safety cases for their continued long-term safe operation. The aims of the TV inspections are to inspect the state of the graphite cores, identify defects and follow the development of previously observed cracks.

The Health and Safety Laboratory (HSL) have developed statistical methodologies for modelling the time to the development of a bore crack in AGRs. Tan and Warren (2009) presented a Weibull model for quantifying the number of cracked bricks at Hinkley Point B and Hunterston B and for forecasting the prevalence of bore cracking in the short-term future, allowing for differences in risk between brick layers and channel locations. Their model, however, did not differentiate between once-cracked and twice-cracked bricks. Although the Weibull model is commonly used in risk analysis, one of its limitations is that it is unable to accommodate non-monotonic risk functions (increasing risk followed by a decreasing risk, or vice versa).

The causes of cracking are not yet fully understood and there is currently uncertainty as to whether the graphite cores have reached stress reversal. Around the point of stress reversal, bore-initiated cracking is anticipated to halt and the development of keyway cracks anticipated to begin. In light of this, Tan and Warren (2011a; 2011b; 2011c) investigated the application of log-logistic distributions to the bore cracking data. The log-logistic distribution, unlike the Weibull distribution, allows an increasing risk function to eventually decline, the timing of the peak being determined by a shape parameter in the model.

In this paper, we present brief descriptions of the most recent statistical models and methodologies that have been developed by HSL for bore cracking, and the results of their application to TV inspection data from HYA and HRA.

THE DATA

The data were obtained from TV inspections at HYA and HRA between 1985 and 2009. At each inspection, the state of the graphite bricks and the presence of defects were recorded. No cracks had been observed in either Layer 1 or 12, hence the exclusion of data from Layers 1 and 12. In our analyses, we define a 'crack' to be a single observed defect in excess of 50 mm in length (defined as a Type III crack by EDF Energy). We assume that a brick can be in one of three physical states:

1) The brick is uncracked.
2) The brick contains a single crack (once-cracked).
3) The brick contains two cracks (twice-cracked).

In our analyses, twice–cracked is taken as the terminating state and no further crack evolution is modelled. As only four bricks containing more than two cracks are present in the dataset, there are insufficient data to model further evolution. This assumption also simplifies the analyses and reduces the (significant) computation time. We also classify the channels according to the number of cardinal neighbours; a channel is classified as a corner, edge or central channel if it has two, three or four, fuel channels as cardinal neighbours, respectively.

The time variable that has been used in the statistical analyses is channel-specific burn-up (GWd) calculated by the EDF Energy PANTHER code. By using channel-specific burn-up as the surrogate for time, we assume that the age of bricks (or the time since start-of-life) within a channel are identical, however the effective channel ages may differ.

STATISTICAL MODELS

In developing statistical models for bore cracking, we assume:

- The data from HYA and HRA can be regarded as coming from the same population and hence can be pooled;
- The cracks are contained within Layers 2 to 11 of the channel;
- At every time $t>0$, the brick is either uncracked, once-cracked or twice-cracked;
- The time, X_1, to occurrence of the first crack can be modelled by a chosen distribution (*e.g.* Weibull, log-logistic) and is a function of time, brick layer and channel location;
- The time, X_2, between occurrence of the first crack and second crack is a function of burn-up alone, and can be modelled by a chosen distribution.

We present models for bore cracking in terms of hazard functions (the instantaneous rate of failure).

The Weibull Model

The Weibull hazard function $h(t)$ is defined for $t > 0$ as:

$$h^{W}(t) = \frac{\beta}{\lambda}\left(\frac{t-\gamma}{\lambda}\right)^{\beta-1} \quad [1]$$

where β is the shape parameter, λ is the scale parameter and γ is the location (or latency) parameter of the Weibull distribution. One approach to incorporating covariates (brick layer and channel location) into the Weibull model is by replacing the scale parameter by a function of the covariates as follows:

$$h_i^{W}(t) = \beta(t-\gamma)^{\beta-1} e^{\theta_i} \quad [2]$$

where θ_i is a linear function of the covariates. The Weibull model, Equation [2], is an example of a *proportional hazards* model, a class of model that can be expressed in the form:

$$h_i^{W}(t) = \phi_i h_o(t) \quad [3]$$

where ϕ_i is a non-negative constant and is the ratio of the hazards of failure at any time for the ith individual relative to a baseline individual (indicated by the subscript 0). The advantage of expressing the Weibull model in terms of the parameters of the proportional hazards model is the ease of parameter interpretation. It can be shown that Equation [2] follows the proportional hazards property, Equation [3]. The proportional hazards model is a convenient general model for comparing the failure times of several groups.

Another approach to incorporating covariates is by representing the Weibull model within an *accelerated failure time* (AFT) modelling framework. The AFT model is an alternative form of survival model in which the explanatory variables are assumed to act multiplicatively on the time scale, that is, the time to failure of the ith case is proportional to that of a baseline case. The hazard function for the Weibull model within the AFT framework case is:

$$h_i(t) = \frac{1}{\sigma}\left[(t-\gamma)e^{\theta_i}\right]^{1/\sigma-1} \quad [4]$$

where σ and γ are AFT model parameters and θ_i is a linear function of the covariates. When incorporating covariates under the Weibull distributional assumption, either the proportional hazards or AFT model may be used.

The Log-Logistic Model

The log-logistic hazard function $h(t)$ is defined for $t > 0$ as:

$$h^{L}(t) = \frac{\frac{\beta}{\lambda}\left(\frac{t-\gamma}{\lambda}\right)^{\beta-1}}{1+\left(\frac{t-\gamma}{\lambda}\right)^{\beta}} \quad [5]$$

where β is the shape parameter, λ is the scale parameter and γ is the location parameter of the log-logistic distribution. One approach to incorporating covariates into the log-logistic model is by replacing the scale parameter by a function of the covariates as follows:

$$h^L(t) = \frac{e^{\theta_i}\beta(t-\gamma)^{\beta-1}}{1+e^{\theta_i}(t-\gamma)^{\beta}} \qquad [6]$$

where θ_i is a linear function of the covariates. The log-logistic model, Equation [6], is an example of a proportional odds model, a class of model that can be expressed in the form:

$$\frac{p_i}{1-p_i} = \rho\frac{p_0}{1-p_0} \qquad [7]$$

where ρ is a non-negative constant and is the odds ratio, the ratio of the odds of survival beyond t at any time for the ith individual relative to a baseline individual (indicated by the subscript 0). The advantage of expressing the log-logistic model in terms of the parameters of the proportional odds model is the ease of parameter interpretation. It can be shown that Equation [6] follows the proportional odds property, Equation [7].

As with the Weibull model, an alternative way of incorporating covariates within the log-logistic model is via the AFT framework. The hazard function for the log-logistic model within the AFT framework case is:

$$h_i(t) = \frac{\sigma^{-1}\left[e^{\theta_i}(t-\gamma)\right]^{1/\sigma-1}}{1+\left[e^{\theta_i}(t-\gamma)^{1/\sigma}\right]} \qquad [8]$$

where σ and γ are AFT model parameters. When incorporating covariates under the log-logistic distributional assumption, either the proportional odds or the AFT model may be used. The log-logistic distribution does not satisfy the proportional hazards property, thus the proportional hazards model may not be used under the log-logistic distributional assumption.

Model Fitting

The Weibull proportional hazards and AFT models and the log-logistic proportional odds and AFT models were fitted within a Bayesian framework. The Metropolis-Hasting algorithm, a Markov Chain Monte Carlo (MCMC) technique, was used for drawing samples from the (joint) posterior distributions of model parameters. The models were fit using MATLAB (The Mathworks, Inc. 2009). Further details can be found in Tan and Warren (2011a).

Model Comparison

When different statistical models are being fitted to data, it may be of interest to compare the performance of the models in terms of model fit. There are several methods of model comparison, one of which is the deviance information criterion (DIC; Spiegelhalter *et al.*, 2002). The DIC is a hierarchical modelling generalisation of the Akaike Information Criterion (AIC; Akaike, 1974). It is applicable to any class of model, can be used to compare non-nested models, and is particularly useful in Bayesian model selection problems. Models are penalised by the model deviance and the effective number of parameters. Spiegelhalter *et al.* (2002) suggest that models receiving DIC within 1 or 2 of the 'best' model are strongly supported, whereas those below 3 have considerably less support. A key advantage of the DIC compared with other techniques is that the DIC can easily be calculated from the output of MCMC analysis.

RESULTS

The proportional hazards model had a substantially lower DIC score than the AFT model under the Weibull assumption. Similarly, the proportional odds model had a substantially lower DIC score than the AFT model under the log-logistic assumption. In light of this, the results from AFT models are not presented in this section; the Weibull results presented in this paper refer to the proportional hazards model, the log-logistic results to the proportional odds model. There was no significant difference between the fits of the Weibull and log-logistic models in terms of DIC. A consistent finding in the models was that the highest risk of cracking occurs in Layer 10 bricks in corner channels, the lowest in layers two to six of central channels.

Hazard Functions

Plots of the hazard function associated with the first crack can be found in Figures 1 and 2 for the Weibull and log-logistic models, respectively. There is considerable uncertainty as to whether the hazard is increasing or decreasing under the Weibull distributional assumption. Under the log-logistic distributional assumption, the hazard is increasing and eventually reaches a turning point, the timing of which is dependent on brick layer and channel location. There is, however, considerable uncertainty in the timing of the turning points.

Under the log-logistic assumption, all forms of the model allowed for an eventual decline in risk, reducing the absolute uncertainty in the hazard function with time. Depending upon the values the model parameters take, this can be a declining hazard (after the end of the latency period) or a hazard that increases before peaking and subsequently declining. The turning points of the hazard functions were calculated for each sampled parameter set and the median and a confidence interval for the turning point of the first crack hazard were calculated from these samples (Table 1). As the current burn-ups at HYA and HRA are about 25 GWd in corner and edge channels, and 30 GWd in central channels, it is unlikely that the hazard functions have already peaked under the log-logistic distributional assumptions.

	Channel location		
Brick layer	Corner	Edge	Central
2 to 6	>100	>100	>100
7	77 (60, >100)	92 (66, >100)	>100
8	60 (49, 77)	71 (55, >100)	87 (66, >100)
9	50 (41, 61)	60 (48, 78)	73 (59, 94)
10	40 (31, 48)	48 (40, 59)	58 (50, 70)
11	61 (50, 81)	73 (56, >100)	89 (67, >100)

TABLE 1: Estimates and confidence intervals for the turning point (GWd) of the log-logistic hazard function, by brick layer and channel location.

The hazard functions associated with the second crack can be found in Figure 3. Due to the limited number of observations relating to second cracks, there is very high uncertainty in the Weibull hazard associated with the second crack soon after occurrence of the first crack (Figure 3a), however the absolute uncertainty decreases with time.

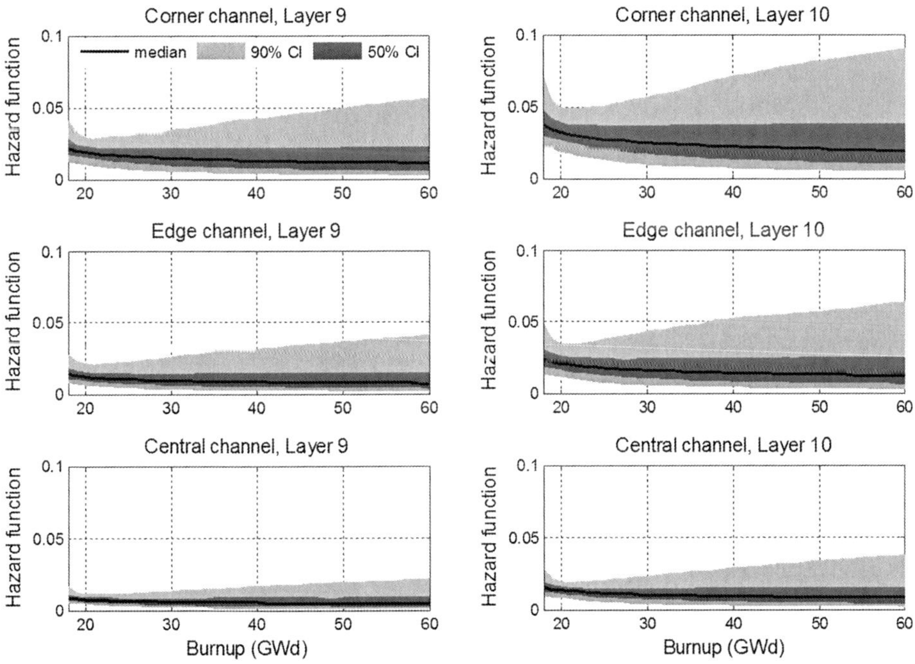

FIGURE 1: The Weibull hazard function associated with the first crack for high risk bricks.

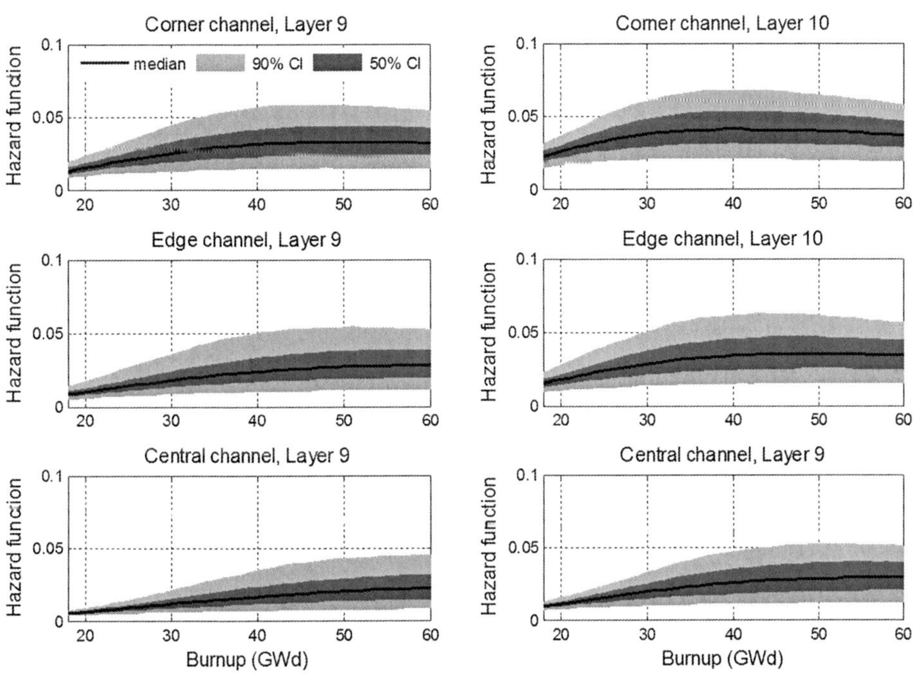

FIGURE 2: The log-logistic hazard function associated with the first crack for high risk bricks.

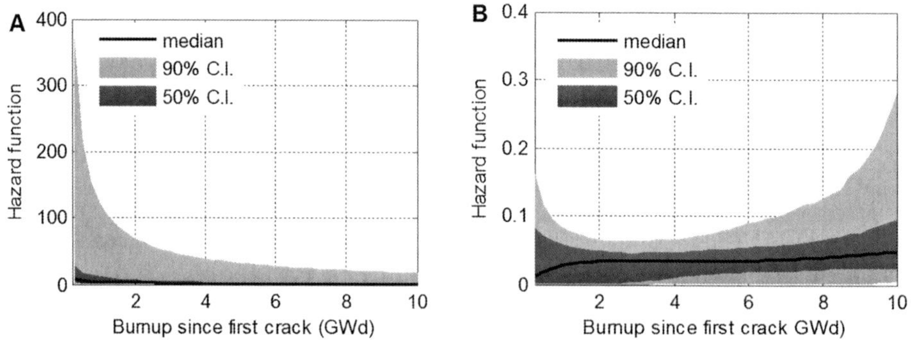

FIGURE 3: (a) The Weibull hazard function associated with the second crack; (b) The log-logistic hazard function associated with the second crack.

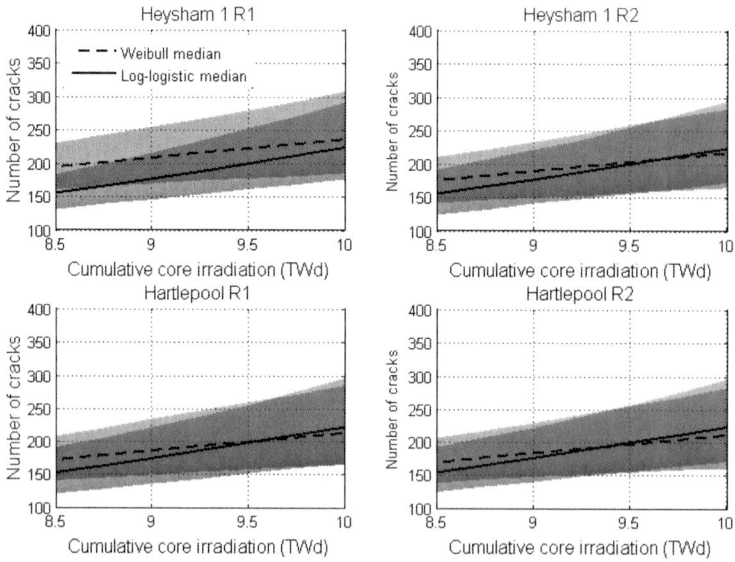

FIGURE 4: Comparison of forecasts under the Weibull and log-logistic models.

Whole Core Forecasts

Whole core forecasts can be obtained under each model by simulating the time taken for each brick to reach states one and two, given the model parameters historical inspection data. Figure 4 shows a comparison of the median forecasts under the Weibull and log-logistic assumptions at each reactor for cumulative core burn-ups between 8.5 and 10.0 GWd; the darker shaded region represents the overlap in the 90% prediction intervals under the Weibull and log-logistic assumptions. It can be seen that that greater numbers of cracks are predicted under the Weibull assumption up to about 9.5 GWd, although the central estimate from log-logistic model forecasts larger numbers of cracked bricks at higher burn-up. However, there is considerable overlap between the 90% confidence regions under the two models.

Under the Weibull model, the greatest number of cracked bricks at a core burn-up of 10.0 GWd is predicted at HYA R1, with a total of 236 (98% C.I. [167, 341]) cracked bricks of which 67 (98% C.I. [36,112]) are expected to be twice-cracked. Under the log-logistic model, the predicted numbers at 10.0 GWd are very similar across all reactors; at HYA R1, 76 (98% C.I. [47,115]) are expected to be twice-cracked out of a total of 224 (98% C.I. [157, 319]).

DISCUSSION

This paper has presented statistical methodology and associated results for quantifying the prevalence of bore cracks at HYA and HRA. A key improvement from earlier work has been the inclusion of parameters representing differences in risk between brick layers and channel locations. The parameters modify some underlying base risk, represented by a probability distribution. There are a variety of methods for modelling differences in risk, and the form of the model conveys differing underlying beliefs about the relative risk. The proportional odds, proportional hazards and AFT models that were examined in this paper are robust with well-understood properties, and their associated parameters have a clear interpretation. The form of the underlying probability distribution restricts how differences in relative risk can be modelled; a proportional odds model is not possible with the Weibull assumption and a proportional hazards model is not possible with the log-logistic assumption. Model comparison using DIC allows the fit of these various models to the inspection data to be compared, and the most appropriate generalisations can be determined.

The use of Weibull and log-logistic models for quantifying the time to cracking has allowed a comparison and quantification of the likely range of risks and whole core forecasts in the short-term future. A consistent finding was that the highest risks of cracking were found to occur in Layer 10 bricks in corner channels, the lowest in layers two to six in central channels. There was little difference between the Weibull and log-logistic models in terms of model fit, however the log-logistic model allows for a more flexible form of the hazard function. There was considerable overlap in the confidence intervals of the short-term forecasts with a total of 236 and 224 cracked bricks predicted under the Weibull and log-logistic models respectively.

There is uncertainty as to whether the risks have been and still are increasing under the Weibull model, with the range of uncertainty increasing as we project into the short-term future. Under the log-logistic model, the absolute uncertainty in the risks of cracking are narrower as we project into the short-term future and there is indication that the risk may be decreasing with time; we cannot exclude the possibility that the hazard functions may already have peaked in the high risk bricks.

References

Akaike, H. (1974). A new look at the statistical model identification, *IEEE Transactions on Automatic Control.* 19, 716-723.

Tan, E. and Warren, N. (2011a). Statistical analysis of bore cracking in Advanced Gas-cooled Reactors: methodology. *HSL Report MSU/2010/34.*

Tan, E. and Warren, N. (2011b). Statistical analysis of bore cracking in Advanced Gas-cooled Reactors: Heysham 1 and Hartlepool. *HSL Report MSU/2011/02.*

Tan, E. and Warren, N. (2011c). Statistical analysis of bore cracking in Advanced Gas-cooled Reactors: Hinkley Point B and Hunterston B. *HSL Report MSU/2011/11.*

Spiegelhalter, D., Best, N. G., Carlin, B. P. and van der Linde, A. (2002). Bayesian measures of model complexity and fit. *J R Statist. Soc B.*, 64, 583-639

Statistical Modelling of Graphite Properties

Peter C. Robinson and Philip R. Maul
Quintessa Ltd., The Hub, 14 Station Road, Henley-on-Thames, Oxfordshire, UK. RG9 1AY
Email: PeterRobinson@quintessa.org

Abstract
At previous conferences organised by The British Carbon Group at Cardiff and Nottingham papers were presented that demonstrated how statistical modelling techniques can be applied to bore cracking of graphite bricks in AGRs. This is now a mature technology with blind predictions being made before each reactor inspection and comparisons with measurements immediately afterwards; the current status of this work is briefly summarised. Statistical modelling techniques have now been applied to the evolution of graphite properties as an integral part of EDF Energy's strategy for modelling thermo-mechanical graphite properties. Examples are given of how such models can contribute to gaining an understanding of key issues for the evolution of reactor cores, supplementing information gained from more complex models based on detailed representations of the underlying processes. It is shown how the representation of variability at different scales is crucial to producing models with useful predictive power for the evolution of the cores. This work is still at a relatively early stage of development, but represents a potentially important contribution to decisions on reactor lifetimes for safe operation.

Keywords
Statistical modelling, bore cracking, variability

INTRODUCTION

During the lifetime of an Advanced Gas-cooled Reactor, the graphite core is subject to neutron irradiation and oxidation. Irradiation directly impacts on carbon atoms; a single carbon atom might be knocked from its lattice position as many as 25 times during the reactor's life. Oxidation through reaction with the CO_2 coolant leads to a loss of mass (generally referred to a weight loss).

Maintaining an understanding of these processes and how they affect the evolution of the graphite bricks is crucial for safe operation of the reactors. An important source of information for this comes from the regular inspections that are undertaken for each reactor. TV inspections are carried out, revealing any cracks that have occurred; physical measurements of bore diameters and channel geometry are made, showing how the bricks have distorted; and samples of graphite are trepanned, allowing direct measurements of properties (including density, dynamic Young's modulus (DYM), coefficient of thermal expansion (CTE), and flexural strength (FS)). Weight loss is inferred from the density and geometry measurements and denoted CMWL (corrected measured weight loss).

The focus of this paper, reflecting Quintessa's role in their work for EDF Energy, is on how the observations made on the operating reactors can be used with statistical models to provide forecasts of their future behaviour. This approach supplements the development of models based on a mechanistic representation of the processes affecting the behaviour of the graphite. Two separate areas are discussed: brick cracking and graphite properties modelling. In each case statistical modelling methods are used to capture the variability and uncertainty inherent in this complex system.

TERMINOLOGY

It is useful to be clear about the terminology that we use as these terms can differ in meaning between different disciplines.

Uncertainty and Variability

In complex systems in general, and graphite cores in particular, it is essential to make a clear distinction between variability and uncertainty:

- Variability is an irreducible characteristic of the system. Thus, it is a real property of a complex system which needs to be understood. Generally, the variability that is of interest here has a largely stochastic nature and the aim is to characterise the statistical properties of the variability (in terms of probability distributions). An example is variation in strength between graphite bricks.

- Uncertainty is a consequence of a lack of knowledge about the system which may be reduced by gathering more data or improving understanding in other ways. Uncertainty is not a property of the system under study, but rather is a property of those who study it. Uncertainty can also be characterised through probability distributions, but these will be subjective in nature as compared to the objective distributions that are sought when considering variability. An example is an incomplete knowledge of how strength depends of porosity structure.

The importance of separating uncertainty and variability arises in part because variability can often be beneficial in reactor safety studies. Variability between bricks means that if new behaviour arises as they age it will not occur in all the bricks at once. Both stochastic variability and systematic variability (*e.g.* between layers of the core) are relevant here. If uncertainty were to be mistaken for variability then this could be non-conservative when developing safety case arguments.

Prediction

We use prediction to mean making a statement about a system, in terms of some measurable quantity, with the intention of comparing this against a future measurement (for the purposes of validation). In general, a single value prediction is not useful; instead a range of possible outcomes is given (a prediction interval), with the range reflecting a level of confidence (*e.g.* we expect the measurement to be within the given interval for 95% of measurements).

Forecast

In contrast to a prediction, a forecast is a statement about a system that is made for use in a safety case or in strategy development. There is no intention that the forecast will be compared against a measurement; indeed it need not be for a measurable quantity. Forecasts will be subject to uncertainty and ranges of outcomes will be given. When using forecasts, conservative percentiles of the possible outcomes are often used (*e.g.* we are 95% confident that the forecast quantity will be lower than a given value).

Blind Prediction

A blind prediction is based purely on information available before a measurement is made. This will include historic data, computer code calculations, and information on the environment where the measurement will be made. Thus, it uses only the information that would be available if a forecast was being made for a safety case. Successful comparison of a blind prediction against a measurement provides strong evidence of predictive power.

Semi-Blind Prediction

A semi-blind prediction additionally uses partial information gathered at the time of measurement. For example, the density of the trepanned graphite may be used to make predictions for other properties, such as DYM. Semi-blind predictions are useful in developing model structures that capture relationships between various properties, but do not provide strong evidence of predictive power on their own.

BRICK CRACKING

Bore Cracks

The status of our work on bore-initiated cracks has been reported at previous meetings (Maul and Robinson, 2005; and, Maul *et al.*, 2008a) and has recently been summarised by Maul *et al.* (2011). In this work, the statistical model that is used relates to the hazard for cracking; this is the probability per unit time for a brick cracking or becoming doubly cracked. The focus is on full length axial cracks, but other crack morphologies, *e.g.* circumferential, can also be considered.

This has now become a relatively routine process that is undertaken on each inspection of any of the Hunterston B/Hinkley Point B (HNB/HPB) and Hartlepool/ Heysham A (HRA/HYA) reactors. At each inspection, a prediction is made for how many new cracks will be seen. This is based on fitting various models to the historic data. Comparisons are then made against the observed number of cracks and the models are refitted and refined as appropriate. There has been a clear trend towards more complex models giving better fits as the amount of data has increased; models with 5 or 6 free parameters now perform best. These models allow for different rates between layers and between central and edge channels. After retuning, forecasts are made for the number of cracks that will be present in the reactor over an extending period (typically about three years) for comparison against core integrity criteria.

Figure 1 shows the predictions and outcomes from some recent inspections. Note that different numbers and types of channels were inspected on each occasion and the predicted ranges reflect this.

Keyway Root Cracks

To date, no keyway root cracks have been seen, but physical models indicate that these are expected to occur late in reactor life. Preparatory work is currently being undertaken using similar statistical modelling techniques to plan what response would be appropriate when/if the first keyway crack is seen.

HYA R2 2008

	D0	D1	D2	D3	D4
S0	72.24	18.74	3.34	0.41	0.03
S1	100.00	32.31	4.59	0.58	0.02
S2	49.43	12.56	2.41	0.15	0.02
S3	26.36	6.36	0.81	0.05	0.00
S4	8.20	1.81	0.10	0.01	0.00
S5	1.28	0.31	0.02	0.00	0.00
S6	0.22	0.04	0.00	0.00	0.00

HYA R1 2009

	D0	D1	D2	D3	D4	D5	D6	D7
S0	6.67	13.33	5.33	3.76	0.94	0.31	0.02	0.02
S1	28.13	36.53	22.56	12.50	4.64	2.14	1.99	0.06
S2	68.48	64.43	42.25	17.84	7.28	2.63	0.86	0.20
S3	77.79	93.64	52.74	32.25	8.65	4.04	1.12	0.23
S4	87.95	100.00	60.39	34.33	11.68	1.56	0.71	0.47
S5	72.80	82.85	49.12	26.15	10.15	2.30	1.42	0.16
S6	45.67	6.46	38.97	16.52	5.75	2.82	0.38	0.08
S7	24.32	30.18	20.93	7.94	3.27	0.58	0.12	0.02
S8	15.35	19.34	14.18	4.34	1.84	0.35	0.52	0.01
S9	10.92	9.39	6.17	1.70	0.43	0.27	0.04	0.00
S10	3.50	4.95	3.04	1.32	0.33	0.04	0.01	0.00
S11	1.03	2.46	1.21	0.22	0.13	0.09	0.00	0.00
S12	0.70	0.65	0.29	0.10	0.02	0.00	0.00	0.00

HRA R2 2009

	D0	D1	D2	D3	D4	D5	D6
S-4					0.00	0.00	0.00
S-3				0.03	0.03	0.02	0.01
S-2			0.51	0.55	0.31	0.12	0.05
S-1		3.69	4.76	2.79	1.31	0.44	0.11
S0	14.51	22.44	17.23	7.30	2.32	0.73	0.20
S1	48.26	52.94	30.14	10.94	4.37	1.12	0.27
S2	77.69	70.67	40.16	15.83	5.14	1.21	0.25
S3	100.00	91.93	36.58	13.28	4.02	1.02	0.19
S4	84.73	63.82	27.40	8.88	2.55	0.60	0.13
S5	57.92	44.13	20.4	6.65	1.69	0.37	0.07
S6	33.10	24.71	9.84	3.36	0.80	0.17	0.04
S7	18.76	12.07	5.60	1.55	0.40	0.08	0.02
S8	8.00	6.10	1.89	0.66	0.16	0.04	0.00
S9	3.07	2.09	0.95	0.22	0.07	0.01	0.00
S10	1.42	0.87	0.34	0.09	0.02	0.00	0.00
S11	0.47	0.29	0.10	0.03	0.01	0.00	0.00

HYA R1 2010

	D0	D1	D2	D3	D4	D5	D6	D7
S-2			0.25	0.39	0.26	0.17	0.05	0.02
S-1		2.92	4.80	3.89	2.36	1.06	0.41	0.11
S0	5.84	17.47	22.42	13.13	7.98	2.72	0.99	0.32
S1	23.78	55.29	49.29	28.99	12.19	5.49	1.55	0.50
S2	40.74	88.5	78.75	8.05	18.64	7.49	2.21	0.61
S3	62.10	100.00	83.62	43.48	19.83	7.06	1.80	0.54
S4	65.73	94.10	74.36	35.44	14.11	5.14	1.23	0.44
S5	52.27	70.02	58.51	27.10	10.45	3.36	0.93	0.21
S6	33.10	46.32	31.04	15.15	6.64	1.93	0.57	0.15
S7	21.07	25.43	16.30	8.54	3.14	1.13	0.30	0.06
S8	9.77	11.29	9.11	4.17	1.33	0.47	0.14	0.03
S9	4.48	6.23	3.62	1.67	0.69	0.18	0.05	0.01
S10	2.07	2.52	1.44	0.74	0.23	0.07	0.02	0.00
S11	0.80	0.86	0.65	0.28	0.09	0.02	0.01	0.00

HRA R1 January 2011

	D0	D1	D2	D3	D4	D5
S0	13.07	8.29	3.64	1.23	0.39	0.07
S1	42.05	31.90	15.56	4.92	1.11	0.27
S2	76.43	56.73	25.96	7.46	2.27	0.52
S3	100.00	83.34	34.98	12.04	3.00	0.63
S4	91.45	69.55	28.90	11.04	2.73	0.57
S5	62.97	51.20	21.21	6.66	2.06	0.47
S6	46.11	38.30	17.04	5.40	1.35	0.33
S7	23.50	18.93	9.13	3.27	0.84	0.19
S8	14.23	10.08	4.06	1.49	0.42	0.09
S9	5.92	4.49	1.86	0.70	0.18	0.05
S10	2.48	1.67	0.91	0.24	0.07	0.02
S11	1.00	0.76	0.36	0.12	0.03	0.01

HRA R1 June 2011

	D0	D1	D2	D3	D4	D5	D6
S0	3.76	4.55	3.54	1.54	0.59	0.21	0.07
S1	15.41	20.34	13.37	6.97	2.47	0.86	0.29
S2	37.88	46.85	35.44	17.64	6.02	1.80	0.50
S3	61.20	88.18	53.64	25.15	9.32	2.68	0.74
S4	78.43	100.00	69.26	29.00	11.48	3.99	0.93
S5	73.78	93.97	65.12	27.01	10.01	3.11	0.80
S6	57.33	83.10	43.62	21.81	7.51	2.28	0.63
S7	40.49	50.12	33.17	14.37	5.59	1.43	0.37
S8	23.43	31.01	18.95	8.67	3.32	0.99	0.27
S9	12.39	16.52	10.73	4.87	1.66	0.43	0.10
S10	6.48	8.07	5.23	1.95	0.69	0.25	0.06
S11	2.89	4.25	2.11	1.13	0.31	0.09	0.02
S12	1.22	1.33	1.06	0.39	0.13	0.03	0.01
S13	0.46	0.54	0.34	0.15	0.05	0.02	0.00

FIGURE 1. Predicted numbers of single and doubly cracked bricks at the six most recent HRA/HYA inspections. In each case the actual outcome is shown. The row and column headings indicate the number of new single and double crack observations expected. The probabilities for each outcome are calculated and then used to order the outcomes; the most probable outcome is scored as 100, with subsequent numbers indicating the percentage of outcomes that are no more probable that the particular outcome. Thus, the zone of numbers above 50 give a set of outcomes that are predicted to occur in 50% of inspections, with the further bands at 95% and 99% so that an outcome in the outer zone is only expected for one inspection in 100.

GRAPHITE PROPERTIES

The work described here on graphite properties has been developed in the context of the CCCA (Core Component Condition Assessment) Validation Protocol (Maul *et al.*, 2008b). The objective in using the protocol is that models used to make forecasts for safety cases should, as far as practical, have demonstrated predictive power.

Iterations of the protocol are associated with trepanning campaigns at the various reactors. The EDF Integrated Methodology models are the main calculation route and so these are the focus of the validation work. In addition, regression models are included and these are the subject of the current paper.

Regression Models

The use of regression models alongside the more complex EDF Energy models has several benefits. First, independent modelling routes of any type are useful in giving indications of model uncertainty and in providing the ability to cross-check and verify results. Regression models have the additional advantage that they are quick to run and it is relatively straightforward to explore important issues and develop new approaches where necessary. When measurement data is compared, some data may be surprising; using regression models allows a range of hypotheses to be explored quickly and efficiently.

Here we look at two examples where methodological developments have been possible in regression models, addressing the important issues of correlations and variability on different scales. Specifically, we look at the correlation structure between FS and DYM and at the separation of variability of weight loss into intra- and inter-brick components.

Correlation

When using variability estimates as inputs to forecasts it is important to capture any significant correlations. Failing to do so can lead to incorrect estimation of variability in key outputs. Failing to account for a positive correlation can lead to over-estimation of variability, which is non-conservative in cases where variability is favourable.

A trial of an approach to handling correlations was used in predictions of the HPB R4 samples taken in 2008. For DYM and FS there are correlations in the central trends because both properties depend on the evolving structure of the graphite. Simple models based on a linear relationship with density perform quite well, as is suggested by scatter plots of the historic data (Figure 2).

In fact the DYM model is better when a term with weighted burnup (W, which includes factors for axial and radial variation in dose) is included and we write

$$DYM = d_1 + d_2\rho_3 + d_3W + \varepsilon_{DYM}$$

$$FS = s_1 + s_2\rho_3 + \varepsilon_{FS} \qquad [1]$$

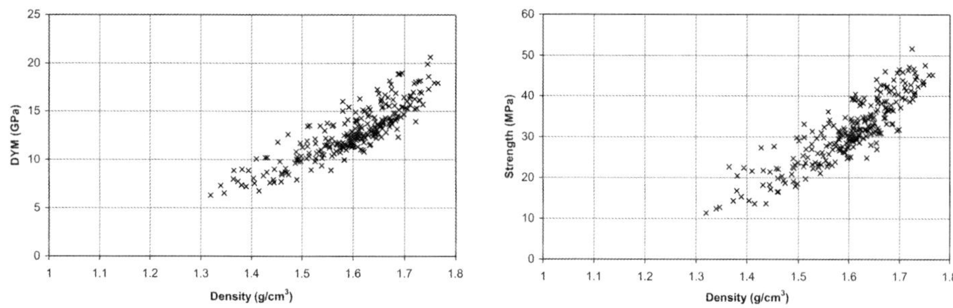

FIGURE 2: Scatter plots for trepanned data from HPB from 2002 onwards, showing (left) DYM versus density (right) FS versus density.

The common dependence on density (here denoted by ρ_3 to indicate the particular density prediction used) makes the central trend correlated, but we are also interested in the correlation between the residuals ε_{DYM} and ε_{FS}. Figure 3 shows the residual correlation for the historic data and the correlation structure that this suggests is a bivariate normal. This becomes the predicted structure for the new trepan samples.

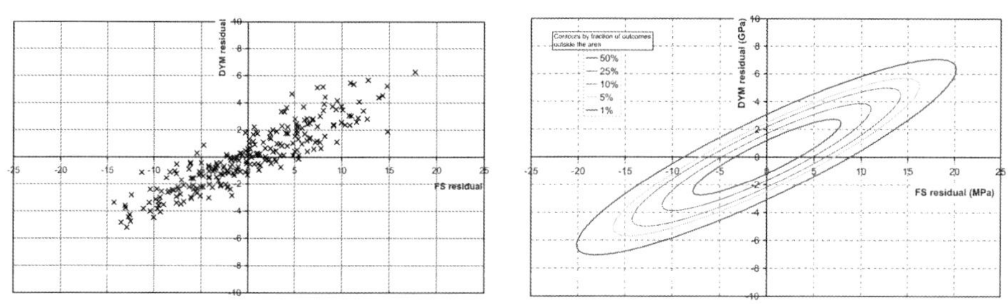

FIGURE 3: Residual structure for HPB DYM and FS data. (left) Residuals after fitting the simple models showing strong correlation. (right) Bivariate normal distribution derived from that residual structure.

The measurements are then compared and the residual structure can be checked. Figure 4 shows the outcome, which matches well, having a correlation coefficient of 0.917 compared to the predicted 0.899.

It is clear that there is a strong correlation structure in the residuals, indicating that the variability for FS and DYM are coupled. The new data continued to show the same structure as the historic data.

Variability on Different Scales

Variability of graphite properties will exist on various scales. Here we focus on the brick-to-brick (inter-brick) scale and within brick (intra-brick) scale. Implicitly, the latter scale is taken to be the scale of the trepan slices used for measuring properties.

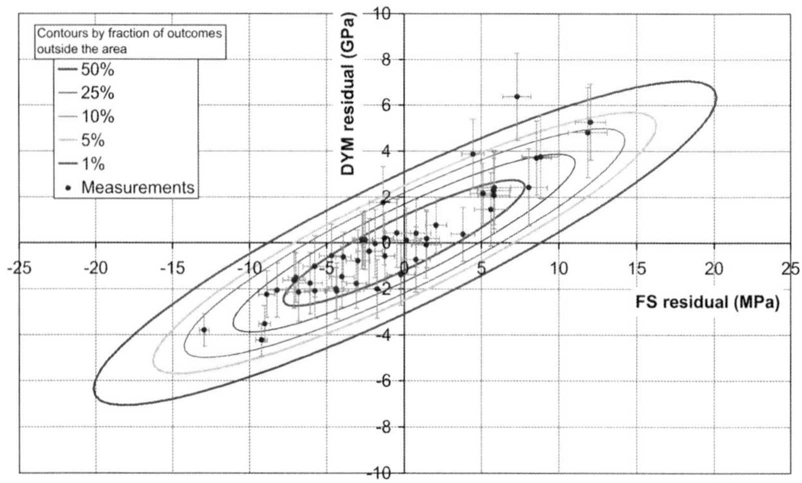

FIGURE 4: Comparison of predicted correlation structure for FS and DYM residuals with the predictions for the HPB R4 trepanning campaign in 2008.

Expert judgement based on the available data is that the brick-to-brick variability dominates. In order to test this quantitatively, we have extended the regression models to represent these two scales of variability explicitly. Here, we give an example of this approach for CMWL. It should be noted that this complicates the analysis and so the first application involved developing a viable approach that could be used routinely in CCCA protocol applications.

This type of modelling has been developed in parallel in several areas of statistical applications and so there is a mixture of terminology in use. The terms hierarchical and multi-level modelling are sometimes used. We are interested in two-level models (brick and slice levels) with the brick level being treated as a "fixed effect". This means that we introduce an extra parameter for each brick that captures the way that it differs from the average. For the purposes of the example given here the fixed effect is simply an offset.

We start with a model that gives a good fit to the historic data. With b denoting a brick and s a trepan slice, the non-linear regression model is:

$$CMWL = \exp\{c_0 + c_1 D_{b,s} + c_1 D_{b,s}^2 + d_3 D_{b,s}^3\} + \varepsilon_{b,s}. \quad [2]$$

A brick-dependent term, u_b, is then added:

$$CMWL = u_b + \exp\{c_0 + c_1 D_{b,s} + c_1 D_{b,s}^2 + d_3 D_{b,s}^3\} + \varepsilon_{b,s}. \quad [3]$$

The brick-to-brick variability, σ_b^2, can be calculated as the variance of the set of u_b values and this can be compared with the local variance, σ_ε^2. Table 1 shows the model comparison. The hierarchical model has an extra degree of freedom for each brick (171 in this example) yet the AIC (Akaike's Information Criterion) is much better. AIC is a measure of goodness of fit which includes a penalty for the number of parameters; a more negative AIC is better.

The variance is seen to be 85% from the brick-to-brick level, confirming the expert judgement.

	Standard Model	Hierarchical Model
Degrees of Freedom	5	176
AIC (goodness of fit)	-5107	-6808
Intra-brick variance σ_{ε}^{2}	1.16 10^{-3}	2.42 10^{-4}
Inter-brick variance σ_{b}^{2}	-	1.38 10^{-3}

TABLE 1: Comparison between standard and hierarchical model fits for CMWL.

Similar results have been found for other properties. Hierarchical models are being included in the current CCCA protocol applications across a number of trepanning campaigns.

CONCLUSIONS

- Variability and uncertainty are unavoidable aspects of modelling complex systems such as an AGR. Statistical modelling techniques provide a way of handling them for various purposes.
- The bore cracking analysis is now routine and has successfully built confidence in the understanding of the rate at which bore cracks develop.
- The use of regression models in graphite property modelling has proved valuable in exploring important issues and provided a framework for developing new methodology.

Acknowledgements

The authors would like to thank EDF Energy for financial support and permission to present this paper. The views expressed in this paper are those of the authors and do not necessarily represent those of the EDF Energy.

References

Maul, P. R. and Robinson, P. C. (2005). A strategy for monitoring bore cracking in AGR graphite cores. *Management of Ageing Processes in Graphite Reactor Cores*, Proceedings of the Meeting held at the University of Cardiff, November 2005.

Maul, P. R., Robinson, P. C. and Steer A. G. (2008a). Understanding AGR graphite brick cracking using physical understanding and statistical modelling. *Securing the Safe Performance of Graphite Reactor Cores*, Proceedings of the Meeting held at the University of Nottingham, November 2008.

Maul, P. R., Robinson, P. C., Suckling P. M., Bradford, M. R., Wheatley, C. J. and Robertson I. (2008b). The development and application of a protocol for the validation of core component condition assessment prediction methods. *Securing the Safe Performance of Graphite Reactor Cores*, Proceedings of the Meeting held at the University of Nottingham, November 2008.

Maul, P. R., Robinson, P. C. and Northrop, P. J. (2011). Statistical modelling of graphite brick cracking in Advanced Gas-cooled Reactors. *Appl. Statist*, 60, [3], 339-353.

Part D

Plant

The Role of Data Visualisation in Core Inspection Decision Making

Claire E. Watson, Peter C. Robinson and Philip R. Maul
Quintessa Ltd., The Hub, 14 Station Road, Henley-on-Thames, Oxfordshire, UK. RG9 1AY
Email: clairewatson@quintessa.org

Abstract
Decisions on which channels to inspect or trepan during a reactor outage require a balance to be struck between many competing requirements for data, as well as other considerations. An important part of making good decisions is having ready access to existing data. Visualising the data on a core map makes it easy to see channels with particular properties. The CHANSELA and PANDA software applications have been developed to support visualisation of the EDF Energy inspection database and PANTHER dosimetry data respectively. CHANSELA also aids the selection of channels by calculating a utility function that balances the various objectives for an inspection. Examples are given of how these pieces of software can assist in decision making by allowing a user to interrogate large amounts of data in an efficient manner.

Keywords
Visualisation, decision making, core inspections

INTRODUCTION

Periodic inspections are carried out on the cores of the UK's fleet of Advanced Gas-cooled Reactors (AGRs) to ensure that no changes have occurred that could threaten safety or economic operation. Various measurements are taken using an array of remotely controlled equipment. The data collected from across the stations has been collated by EDF Energy in a database, and is used to inform safety case decisions, as input to computer simulations of various aspects of the core, and to make decisions about how many and which channels to inspect in the future.

There is, therefore, a large amount of data available regarding the state of the core; that collected at inspections (of which there is a limited supply), and that produced as output from computer simulations (of which there may be a great deal). The information contained within this data ultimately informs safety cases and controls the lifetime of the reactors and it is therefore imperative that it is analysed in an effective and coherent manner.

Data visualisation is one way of disseminating this information. Not only is a graphical representation of data easy to understand by all, but the human eye is remarkably good at picking up patterns and trends. Anomalies that may not be otherwise obvious can spring out from the page if data is visualised in an effective way. The most natural way to display data concerning a nuclear reactor is to use a core map, a top-down view of the fuel channels between the graphite bricks (see Figure 1 for an example). In this paper we present two pieces of software, CHANSELA and PANDA, which have been developed by Quintessa for EDF Energy and which employ core maps and other visualisation tools to assist users in the analysis of core data.

CHANSELA

Channel Selection

It is not feasible to inspect all the channels in the core when carrying out an inspection, since the reactor must be offline whilst the inspection tools are in place and this is a very expensive undertaking. Typically inspections occur during statutory outages, once every two or three years, which last for approximately 20 days. During this time, it is usually possible to inspect about 30 channels (a typical core contains around 300 channels), although at some stations physical constraints such as the lack of a buffer store restrict this number. In order to choose which channels to inspect during an outage, a balance must be struck between competing aspects which include: the need for information to be collected; the requirements of the relevant safety case; the cost of inspecting each channel; physical constraints (*e.g.* of the equipment used to carry out the inspections, or of the reactor structure); and any other issues that may have an impact on the decision, for example, the refuelling schedule of the station.

Making a good decision relies heavily on having ready access to existing data. This data may be a list of channels to be refuelled, knowledge of which channels are difficult to access with the inspection equipment, measurements taken from previous core inspections, properties calculated from these measurements or by other means, *etc.* In the past, this data was often collated by hand from numerous sources and decisions made via group discussions using the data as input. Without an effective way to visualise the large amounts of data, this process is very time consuming and difficult.

With this in mind, the software tool CHANSELA (CHANnel SELection Assistant) has been developed to assist in the process of selecting channels for inspection. This graphical application interfaces with the inspection database collated by EDF Energy, which contains the measurements taken at (or data inferred from) each core inspection such as channel bow, shrinkage, ovality and crack descriptions.

The primary function of CHANSELA is channel selection, based on user-input data requirements, constraints and preferences. The software can either score a portfolio of channels selected for inspection by the user, or it can automatically find an optimal portfolio of channels using a genetic algorithm. This use of CHANSELA was described at the 11th INGSM conference held at Eastbourne in 2010, but here we concentrate on a secondary function of the software, which is visualisation of the data which serves as input to the channel selection decision process.

Data Visualisation

Data are displayed in the software on a core map. Use of this data is not restricted to the channel selection process; it may be used as input to model development, inspected for patterns and trends, and many other uses. Each piece of data is termed a "characteristic" and the software has a Characteristic Viewer mode (Figure 1) which displays the selected characteristic on a core map. If multiple inspections have been carried out on a particular channel, only the most recent data is shown (this is relative to the date of the inspection campaign, which can be set retrospectively in order to view historical data). In the core map displayed in Figure 1 the colour of each channel represents the number of TV inspections that have been carried out on that channel, as indicated by the legend to the right.

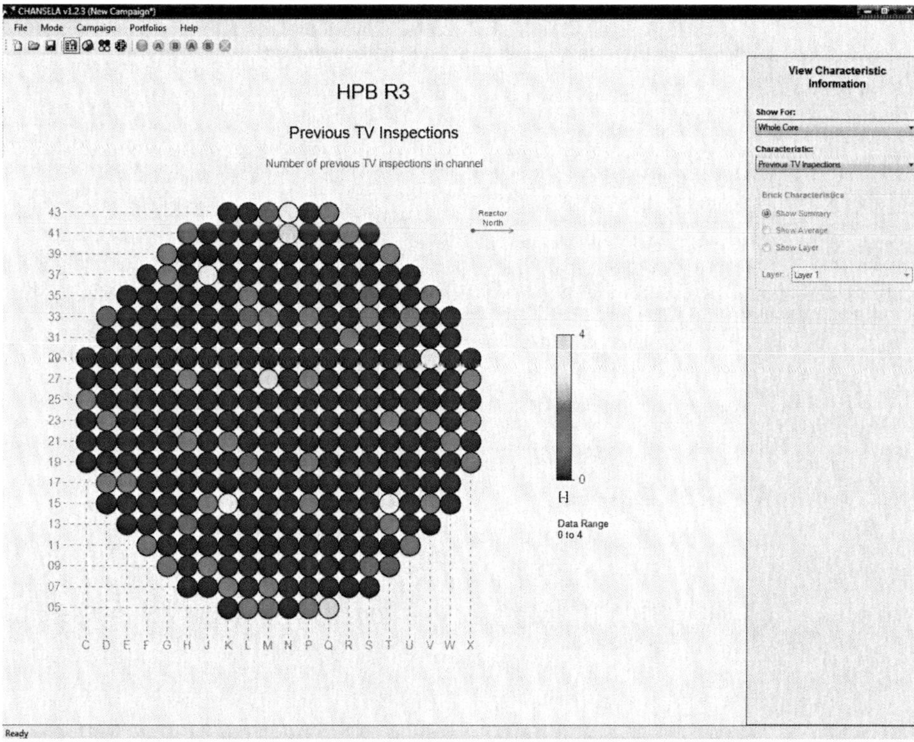

FIGURE 1: CHANSELA displaying the number of previous TV inspections carried out in each channel (for Hinkley Point Reactor 3).

Some data may relate to individual bricks within a channel, for example crack descriptions or ovality. In that case, there are a number of different ways that CHANSELA can be used to display the data: for each individual layer (Figure 2); a summary of all the layers (Figure 3), where each channel is divided into a cartwheel with each segment representing a separate brick (proceeding in a clockwise direction starting with layer 1 at 12 o'clock); or an average value for the channel over all the bricks (Figure 4).

CHANSELA does not only deal with numerical data but also that formatted as dates (*e.g.* the date of the last inspection), Booleans (true/false, *e.g.* channels considered high risk) and data split into categories (*e.g.* crack classifications), as shown in Figure 2.

The data available in CHANSELA can be extended beyond that contained within the inspection database. Some data may be relevant to channel selection, *e.g.* channel irradiation, which is not measured but is calculated. Other data may only be relevant to a particular outage or a particular station and is provided by other means. These kinds of data are termed "extra characteristics" and can be imported into a CHANSELA campaign if they have been formatted in the correct style. This flexibility means that the capabilities of the application are easily extended without the need for any software engineering.

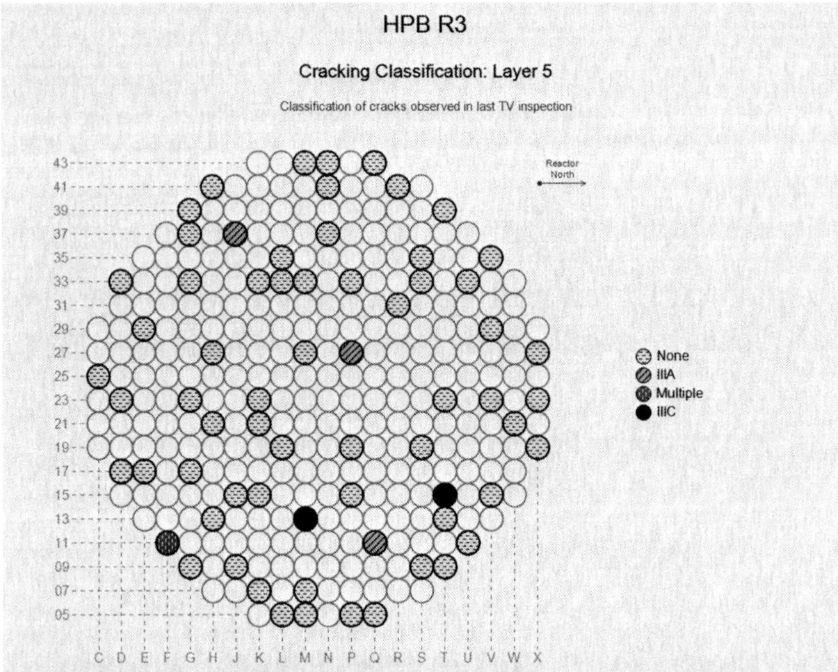

FIGURE 2: A brick characteristic (cracking classification) displayed for layer 5. This is a brick characteristic that is split into multiple categories. Blank channels have not been inspected for cracks.

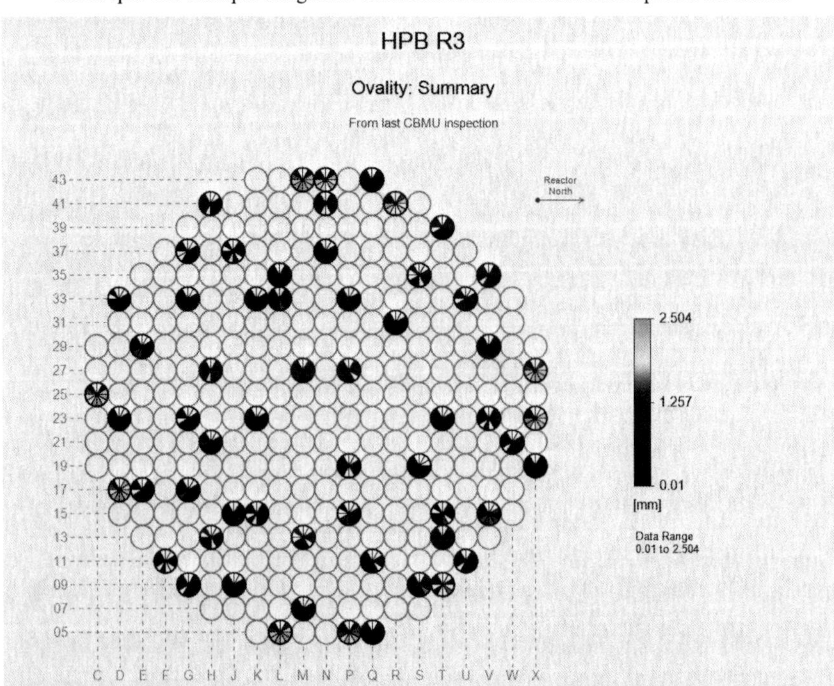

FIGURE 3: A brick characteristic (ovality) displayed for all layers. Each segment in the cartwheels represents a layer, with layer 1 at 12 o'clock and proceeding in a clockwise direction.

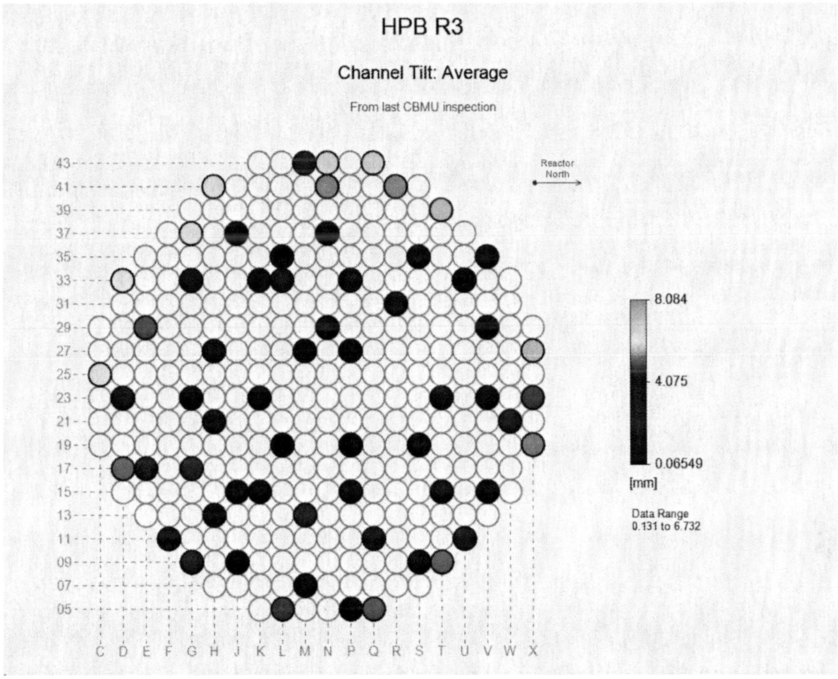

FIGURE 4: A brick characteristic (channel tilt) displayed as the average over all the layers in a channel.

Rules

An extremely useful part of the software is the ability to combine characteristics into "rules". These rules are ultimately used to provide the criteria for channel selection, but are also a means for picking out certain channels of interest. A standard set of useful rules is provided automatically but the user is able to extend these to cater for the needs of the particular inspection campaign being considered.

Rules can use simple logical operators such as >, < and !=. For example, a simple rule may be constructed as follows to locate channels with a large channel bow:

Maximum Channel Bow > 3 [mm]

Alternatively, if the set of channels with the top 10% of shrinkage measurements are required, one could add a rule

Maximum Shrinkage HIGH 10%

It is also possible to find the bottom 20%, or the top or bottom *N* where *N* is any integer. By combining rules with AND, OR and NOT logical operators it is possible to construct complex rules. For example, the rule

Maximum Channel Bow > 3 [mm]
AND Maximum Shrinkage HIGH 10%
AND NOT Cracking Classification = None

would pick out channels with a large channel bow and shrinkage that contain cracks (note that if a brick or channel has no data, *i.e.* in this case has not been inspected for cracks, it is assumed that it does not meet a rule).

Recent additions to the rule capability include the ability to select channels that are adjacent to others that may meet a rule (*e.g.* channels neighbouring high shrinkage channels), and the ability to specify ranges (*e.g.* channels inspected more than twice between 2005 and 2009). This creates a flexible tool for interrogating large amounts of data.

PANDA

The data visualisation capabilities of CHANSELA have proven to be so useful that a separate tool, PANDA, has been developed to visualise data produced by the PANTHER code (Hutt *et al.*, 1991), which calculates various core and fuel element properties such as fuel element burnup and channel power.

The dosimetry data is provided in the form of formatted text files. Each file contains output for a single reactor, across each channel (and for some forms of data for each fuel element in a stringer), for multiple "snapshots" in time (on approximately a weekly basis in latter years). Thus there is a very large amount of data available which needs to be disseminated.

The primary visualisation tool in PANDA is the core map, which is similar to the CHANSELA characteristic viewer. Thus fuel element data (e.g. element burnup and irradiation) can be displayed in a number of ways; for a specific element, as a summary over all elements (using a cartwheel graphic similar to that shown in Figure 3 for CHANSELA), or, in appropriate cases, as the total for the channel summed over all elements. An example is shown in Figure 5. This example is a single snapshot, but the user is also able to plot incremental data by selecting two snapshot dates. The software then shows how the selected data has increased or decreased over this period on the core map.

In addition to the core map display, PANDA has the added option of displaying a time history for particular channels or fuel elements. This can be plotted against either the total core burnup (Figure 6a) or against the data (Figure 6b). The raw data can be exported from PANDA in CSV (comma separated value) format, to be manipulated further in Excel or some other tool. This is true of both time history and snapshot (core map) data.

Data Management

An important aspect of data use is the knowledge that the data has not become corrupted or been edited either purposefully or accidentally. The dosimetry data produced by PANTHER is contained in human readable text-based files. PANDA takes two steps towards ensuring that this data is protected. Firstly, when a new PANTHER data file is introduced, PANDA converts it to a binary format which is not easily edited manually. Checksums are used to ensure that the binary files are indeed unaltered; these are akin to an electronic fingerprint that is unique to a particular file containing specific data. Secondly, whenever new PANTHER data is available and introduced to PANDA, a check is made to ensure that any historic data matches that already stored in the binary file. These two aspects ensure that users can be confident the data accessed by PANDA is the same as the raw data output by PANTHER.

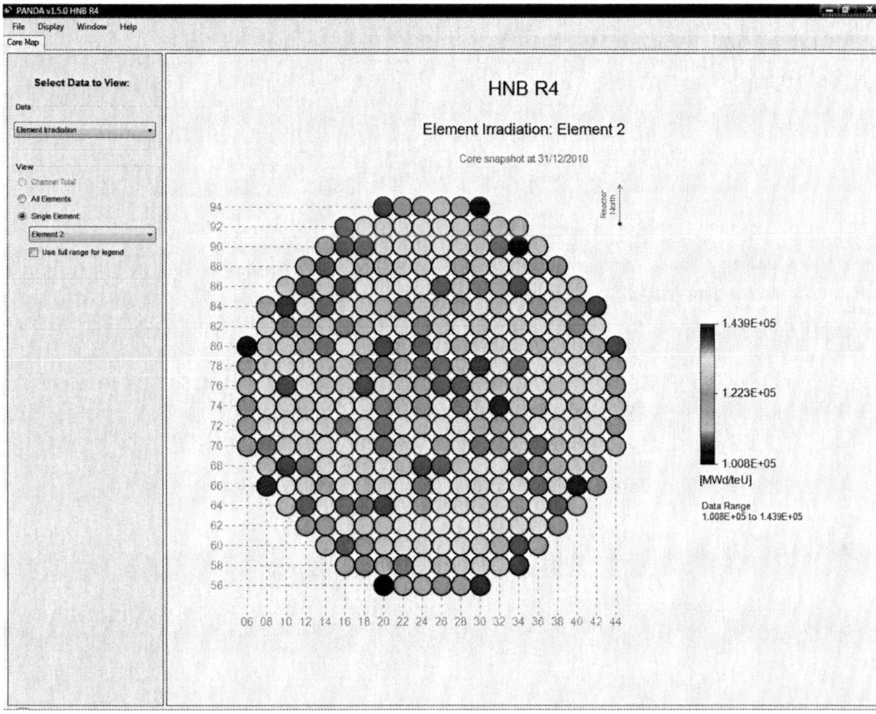

FIGURE 5: PANDA displaying irradiation for element 2, for Hunterston B Reactor 4.

Planned Developments

Version 2.0.0 of PANDA is due for release at the end of year. The ultimate aim for the software is to provide a unified approach to the management of PANTHER data, which is currently used by a number of groups throughout EDF Energy and beyond. To this end, the output CSV files will be made fully configurable thus allowing users to extract the data in the format that they require, possibly suitable for direct incorporation into other codes. This helps to reduce copy and paste and other errors that inevitably creep in when data is handled manually.

The current version of the software displays the number of times fuel in a channel has previously been shuffled, and this will be augmented in version 2.0.0 with a full history of where the fuel has been before.

Finally, new methods of viewing data will be added for each snapshot. The user will be able to view data for a snapshot either as a histogram (with definable categories) or as a CDF (cumulative distribution function). The user will be able to select a subset of channels to appear in these plots, based on the rings that they occupy in the core. As with the core map and time history data this can be exported to a customisable CSV file for use in other applications if desired.

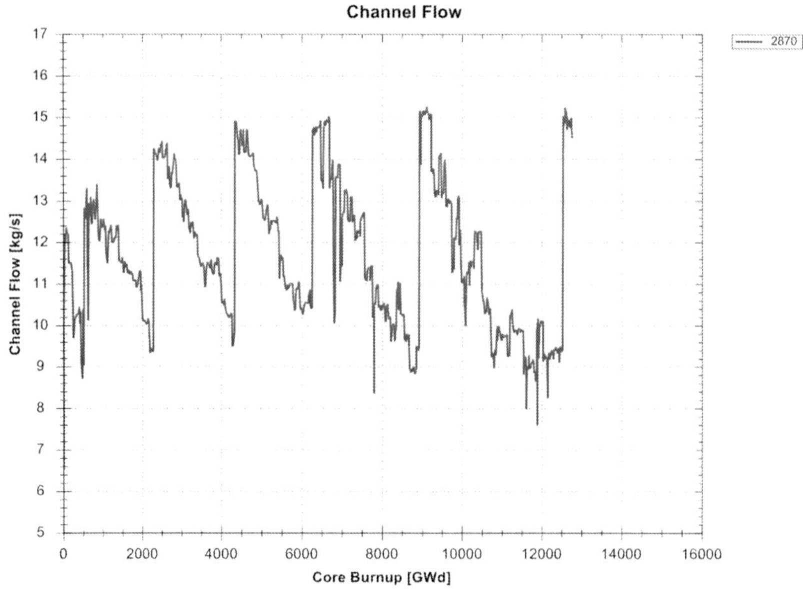

(a)

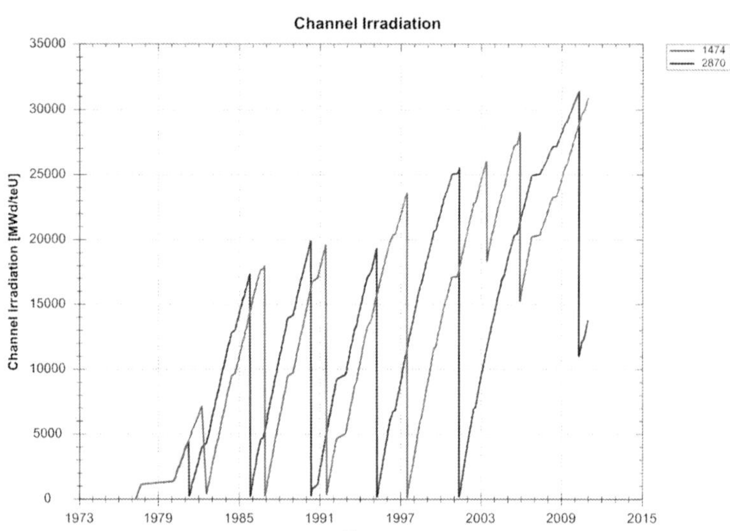

(b)

FIGURE 6: Time history output from PANDA with (a) Channel flow vs. core burnup; and, (b) Channel irradiation vs. date

Link with CHANSELA

As mentioned previously, extra characteristics can be imported into a CHANSELA campaign if they are presented in the correct data format. PANDA has a built-in capability to convert PANTHER output to the CHANSELA extra characteristic format, thereby allowing new characteristics such as channel irradiation and gas outlet temperatures to be included in the decision-making process.

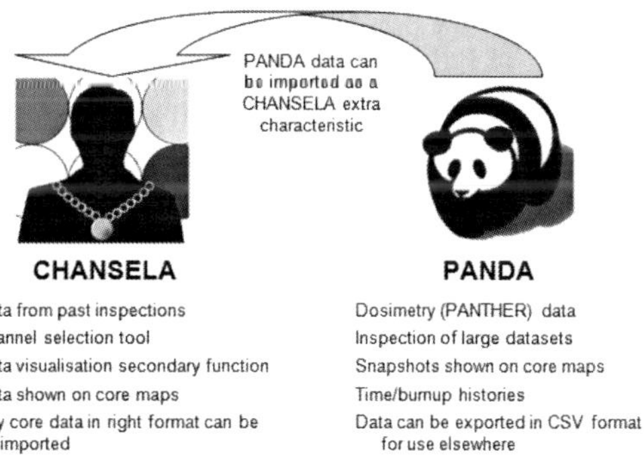

FIGURE 7: A summary of CHANSELA and PANDA and the link between the two.

CONCLUSIONS

Data visualisation is an immediate and effective way of disseminating and analysing large quantities of data. By displaying information about the core status on core maps, CHANSELA and PANDA provide invaluable support to those making complex and difficult decisions about channel inspections. CHANSELA takes this one step further by employing this data to meet specified requirements when automating the channel selection process. The use of tools such as these along with the development of new ways of visualising core data makes the decision process easier, which can only be a good thing in terms of both safety and economic considerations.

Acknowledgements
The authors would like to thank EDF Energy for financial support and permission to present this paper. The views expressed in this paper are those of the authors and do not necessarily represent those of the EDF Energy.

References
Hutt, P. K., McEllin, M., Halsall, M. J., Gaines, N. and White R. J. (1991). UK core performance code package. *Nuclear Energy* 30, 291-298.

Non-Linear Seismic Assessment of AGRs: Incorporation of Aged Graphite Material Properties

Kieran Scully and David Knowles
Atkins, The Hub, 500 Park Avenue, Aztec West, Bristol BS32 4RZ, UK
Email: kieran.scully@atkinsglobal.com

Abstract

A requirement exists to demonstrate that the Advanced Gas-cooled Reactors (AGR) provide a single line of protection against a seismic event of defined probability. In order to demonstrate this requirement, a 3D non-linear explicit modelling methodology of the graphite bricks and supporting structures has been developed. The modelling methodology, developed in LS-DYNA, is used to predict component loads, velocities, displacements and control rod / fuel stringer channel shapes during and post seismic event. This paper describes the modelling process with which detailed graphite brick shrinkage and weight loss predictions are incorporated into a global seismic model of the reactor core.

Keywords

Seismic, non-linear modelling, graphite irradiation

INTRODUCTION

The graphite bricks in the AGR cores are arranged into an array and stacked over a number of layers. The brick array consists of both moderator (fuel) and interstitial bricks which interact through a system of keys and keyways – see Figure 1. The key/keyway system maintains the position of bricks relative to each other and allows thermal expansion of the core. The layers of bricks form channels through which the fuel and control rods can be inserted.

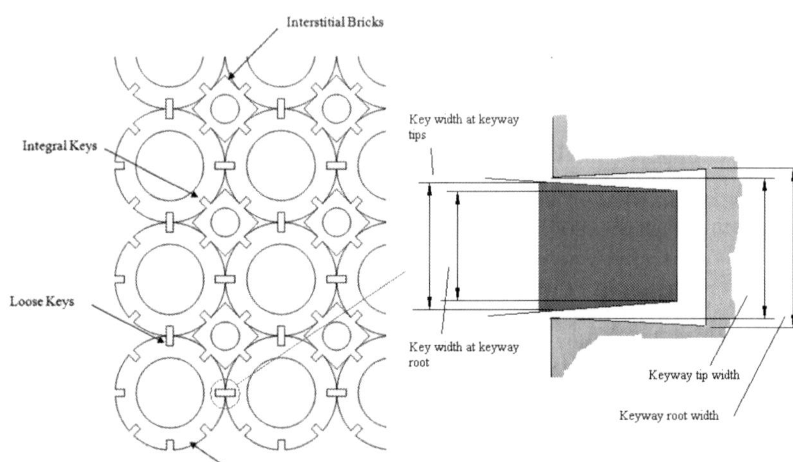

FIGURE 1: Illustrates graphite brick array, loose and integral keys and key engagement.

A non-linear dynamic model (GCORE) has been developed by Atkins to assess the seismic capabilities of the AGR reactors. Discrete nodal rigid body models are used to represent individual graphite bricks. Interactions between bricks and the keying system are modelled using non-linear spring and damper pairs. This paper presents the methodology for modelling interactions between graphite bricks and the implementation of brick dimensional changes and degradation effects due to the effects of irradiation, for a number of contact cases.

OBJECTIVES OF IRRADIATION MODELLING

The objectives of modelling brick degradation are to demonstrate that the graphite ageing mechanisms do not adversely affect the global seismic capabilities of the AGR reactors. The primary degradation mechanisms represented in the seismic assessments are:

- Brick direct contact - increases in the contact clearance between horizontally adjacent bricks occur due to radial shrinkage.
- Keyway pinching - reduces clearances in the keying system. Shrinkage rates at the brick bore are greater than at the periphery, resulting in dovetailing of the keyway. The ultimate effect of this is termed keyway pinching, where, after a number of operational years the key becomes pinched at the keyway tips – see Figure 1 for terminology.
- Weight loss of the bricks - affects the inertial properties of the bricks.
- Loss of material strength - affects the structural integrity of the graphite bricks. Predominantly the seismic assessments consider loads on the keying system.

The seismic model is generated through an Excel based Visual Basic Application (VBA) program which contains the required material and dynamic properties to represent the core at chosen particular ages (measured in full power years –fpy).

SEISMIC GRAPHITE BRICK MODELLING[1]

The graphite bricks are represented by nodal rigid bodies with nodes placed at interaction points (*e.g.* key/keyway interaction and direct contact points). The nodes are distributed over a number of (nodal) layers and can be modified to represent different brick design features if required. The contact points and nodal layers, of which there are five per moderator brick and two per interstitial brick, as indicated by the scenario in Figure 2. The contact types incorporated as part of the seismic model include the following:

- Direct contact – between horizontally adjacent bricks.
- Shear contact – represents a translational clearance determined by the extents of key and keyway engagement.
- Longitudinal contact – Compression of a key between two cracked bricks (not considered further as part of this paper).
- Vertical contact – represents the brick rocker motion (not considered further as part of this paper).

[1] Termed GCore brick model in this paper.

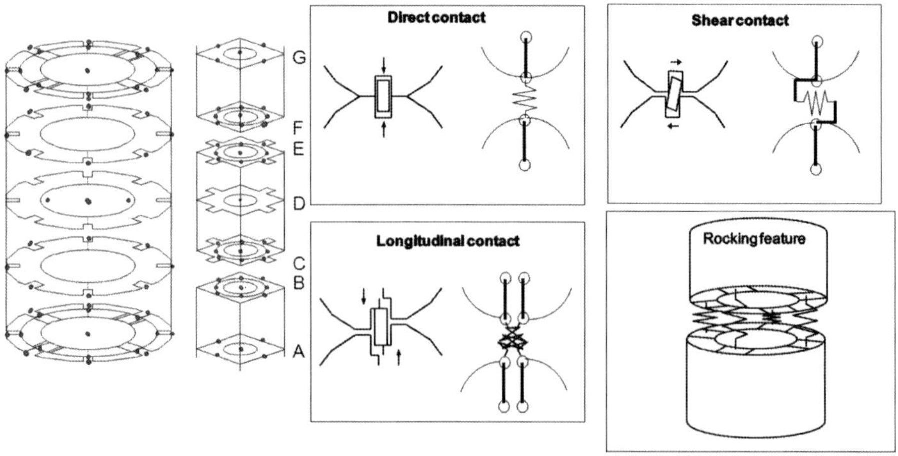

FIGURE 2: Illustrates brick rigid body nodes and contact types.

It should be noted that interstitial contacts are not shown in Figure 2, but are modelled in a similar manner. Contact between the interaction nodes is modelled using non-linear spring/damper pairs. Clearance between keys and keyways (termed lost motion) is represented in the contact spring curve by the inclusion of a spring extension with zero resistance. Subsequently, at the point of contact a linear force / displacement curve is implemented (see Figure 3). Similarly this method is applied to the appropriate damper. The damper is defined using two inputs, a force/velocity curve and a scale factor curve (which includes the contact clearance gap). The majority of brick to brick contacts in the core are defined using these non-linear curves. For the purposes of this paper, only the assumptions made when calculating the direct contact and shear contact are considered.

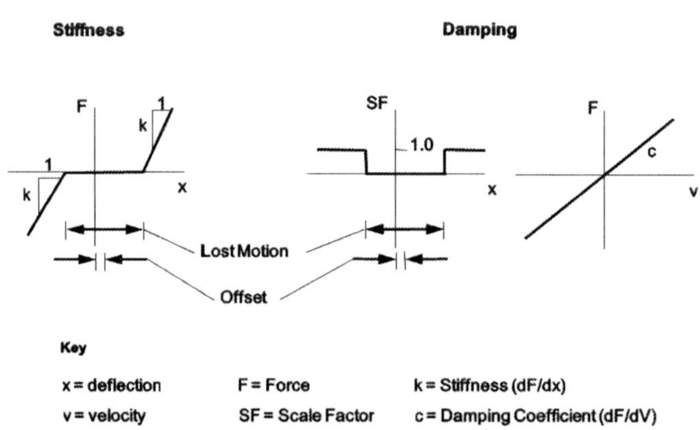

FIGURE 3: Non-linear contact stiffness and damping curves.

GRAPHITE BRICK DIMENSIONAL CHANGES

The detailed graphite brick, loose and integral key dimensional and material properties changes for each station are predicted using a 1/8th symmetrical finite element model. The model (developed by Serco, see Slater (2009)) incorporates the evolution of graphite material properties and internal stresses for a peak rated brick throughout reactor life. An extensive range of data is calculated for a number of material related parameters at intervals of 2 fpy until the expected end of life (40 fpy). This includes a period of shrinkage followed by expansion (termed "stress reversal") towards the end of brick life. Changes in the graphite brick shape are output as coordinates for each available node.

A number of modelling assumptions need to be made to incorporate the detailed Serco brick model data into the relatively few contact interactions in the seismic model. The following section provides details on the assumptions made in interpreting the shrinkage data.

Direct Clearance

The direct clearance is defined as the contact gap between two horizontally adjacent bricks. Assuming the distance between brick centre axes (or brick pitch) remains constant the change in horizontal direct clearance between bricks is directly related to the radial brick shrinkage. The seismic model defines direct contact between both horizontally adjacent moderator bricks and interstitial bricks. These contact clearance values vary across the core with brick type and irradiation dose – as described below.

Predictions of graphite shrinkage conclude that radial shrinkage varies along the brick axes by the order of 1 – 2 mm throughout the brick lifetime. The calculation of the irradiated brick radius incorporates the radial shrinkage of the keyway tip nodes at the top, middle and bottom of the brick. The mean value is subsequently calculated for use in the seismic model. This assumption has been made as there are limited direct contact interaction nodes in the seismic model and in addition, direct use of the maximum or minimum shrinkage would have a cumulative effect on the seismic behaviour over the entire core. The shrinkage of the loose and integral keyway tips occurs in the radial and tangential directions and both are incorporated in the calculation – see Figure 4. In addition thermal expansion of the brick is also taken into account. The final direct clearance between adjacent bricks is equivalent to the thermally expanded brick pitch minus the adjacent brick mean irradiated radii.

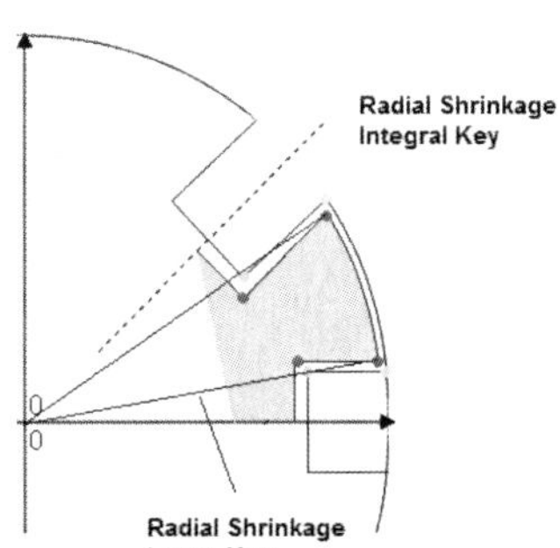

FIGURE 4: Radial shrinkage of loose and integral keyway tips.

Shear Contact (Lost Motion Calculation)

The shear contact clearance takes into account a number of changes in brick geometry, namely key/keyway widths at the tip and root and the key /keyway depths between opposing keyways. These parameters can subsequently be used to determine the translational and rotational clearance (which when combined is termed the lost motion) of the key in the keyway – see Figure 5.

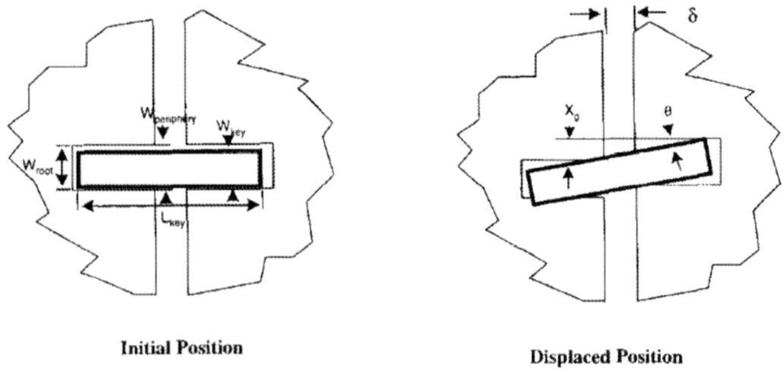

FIGURE 5: Illustration of lost motion, x_g.

The keyway width at the root and tip vary along the vertical axis of the brick. With a limited number of key/keyway interaction nodes (two per key) in the GCORE brick model it is assumed that the keyway tip and root widths are equal to those at the midpoint along the height of the loose or integral key, *i.e.* the data is extracted at the keyway nodes closest to the vertical midpoint of the loose or integral key.

This assumption is justified to properly capture the effects of pinching in the calculation of the lost motion. Keyway shrinkage is impaired through contact with the key (post keyway pinching) and if an average width is taken over the height of the keyway the extents and timings of pinching are overestimated. This assumption has been correlated with the Serco brick model.

The key widths in the clearance calculations are taken directly from the Serco model at the tip and symmetry plane. Shrinkage at the key tips is slightly greater than at the symmetry plane – as illustrated by Figure 1.

Subsequently, the translational and rotational clearances are summed and can be modelled as a single shear clearance. This assumption can be made as generally the clearances remain small and sustained large rotations between bricks do not occur during the assessments.

BRICK SHRINKAGE AND THE GLOBAL SEISMIC MODEL

The global seismic model incorporates a discrete representation for each brick within a half core. Currently detailed predictions of brick shrinkage are available for peak rated bricks. A number of modelling assumptions are made about the irradiation dose received in regions extending both radially and axially from the core centre including the interstitial bricks. The irradiation dose is reduced by a form factor which is calculated as a fraction of the peak rated brick dose, typically this is quoted as a normalised value (relative to the peak rated brick). Radially the core is divided into three groups consisting of the inner core, inner periphery and outer periphery rings (the latter two are collectively termed the reflector region). The reflector region has a radial form factor of zero and hence is assumed to receive no irradiation dose, so only thermal expansion is considered.

Axially the bricks are divided into four layer groups, in which the number of brick layers per group varies between station designs. Typically the lower most and upper most brick layers (extreme top and bottom of core) are considered to receive no irradiation and the two layer groups in between are attributed the appropriate axial form factor.

The process of extracting the shrinkage data remains as described above however values are taken at the appropriate core age. Similarly shrinkage values for interstitial bricks within the inner core are based on 2/3 of the dose for the peak rated brick. For interstitial bricks above and below the inner core region the axial form factor and 2/3 dose are both applied reducing the effects of irradiation further. As per the moderator region interstitial bricks in the reflector region receive no dose.

BRICK MASSES AND WEIGHT LOSS

The radial and axial layer groups are also used to incorporate different brick masses in the model. In particular moderator bricks and reflector bricks, the latter of which are solid in construction. In addition the radial and axial layers groups also allow the functionality to appropriately model brick weight loss of these zones. As per brick shrinkage, mass reduction decreases both axially and radially away from the inner core and the form factors are utilised to calculate the appropriate brick dose (and weight loss) for that region.

INCORPORATING LOSS OF MATERIAL STRENGTH AND SEISMICALLY-INDUCED DAMAGE

The current GCORE model does not incorporate real time damage mechanisms. Assessment of the keying system is conducted retrospectively using capacities calculated from physical brick testing appropriately modified for:

- Changes in strength due to irradiation – incorporates the affects of fast neutron irradiation and weight loss caused by radiolytic oxidation.
- Fractional remnant strength (FRS also known as ΔS) – reflects the combination of the interactions between internal stresses. This mechanism predominantly affects external load bearing capacities after stress reversal.
- Changes in key/keyway prying forces caused by shrinkage.

Any predicted failures are included as part of a subsequent modelling iteration through the removal of the spring / damper pairs associated with the relevant key.

Component Strength and the Effects of Irradiation

Destructive brick testing was undertaken as part of the design substantiation for each type of reactor core. A number of different scenarios and loading configurations were carried to determine fracture loads of both the keys and the keyways. The results of the tests, including the strength reduction mechanisms mentioned above, are used to generate the key/keyway capacities for post processing.

The effect of irradiation on the component strength is determined using the equations and method presented in Data sheet C2/2 of P28 (McLachlan (2003)). The ratio of the irradiated and un-irradiated strength terms (σ/σ_0, calculated using the pinning, structure and weight loss terms) is available directly from the Serco brick model from any nodal position[2]. However,

[2] It should be noted that the ratio of σ/σ_0 is greater than 1 for the majority of brick life.

irradiation effects are not constant across the keys/ keyways and the following assumptions have been made in the data extraction:

- The σ/σ_0 values for the fuel brick loose and integral keyways are taken as the average of those at the respective keyway root corner. The average of the top, middle and bottom of the keyway is taken. The σ/σ_0 values at the keyway root have been taken as this location is a significant stress raiser and the crack initiation site under external loading. The variation of the strength term over the height of the keyway is negligible and an average incorporates these small variances.
- The σ/σ_0 values for the integral keys and loose keys are taken as those at the relevant fuel brick keyway tips. Again this is an average value across the length of the keyway.

The Fractional Remnant Strength, as described earlier, represents the effects on component strength from the interaction between stresses from different loading conditions. This mechanism predominantly affects the external load bearing capacities after the onset of stress reversal and varies significantly along the length of the keyway. Strength reduction is greater at either ends of the brick and so it would be overly conservative to use an average value over the entire height. To that end the average over the height of the loose and integral keys has been taken. It should be noted that the FRS term does not apply to the keys themselves as the degradation mechanism results from the interaction of internal stresses across the brick section caused by its complex geometry. The FRS data is readily available from the Serco brick model.

Additional consideration is also given to the increased prying effects of the keys in the keyways based on the changing geometry. As radial shrinkage of bricks increases the position of the key relative to the keyways has the potential to generate increasing prying effects on the keyways. This has been illustrated as a three point contact situation in Figure 6. As the diagram illustrates increasing brick to brick clearance (labelled x_1) will decrease the lever arm x_2 should the key remain in the same engaged position. Assuming a 3-point contact situation the acting load C is equal to the ratio of B and A.

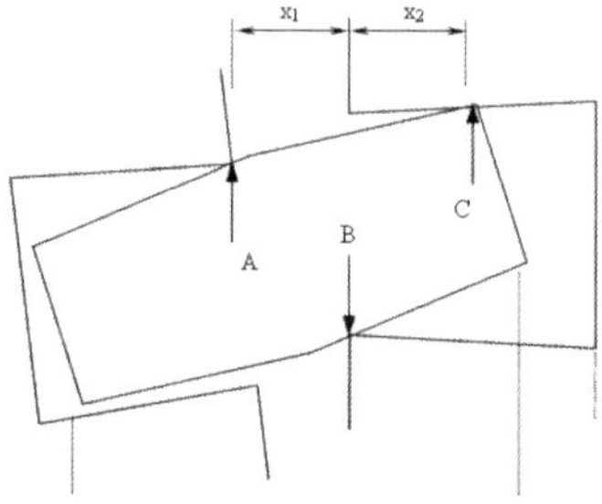

FIGURE 6: Illustration of increased keyway prying loads through changing brick geometry.

This ratio is calculated using the values extracted from the Serco brick model as described above. The method is considered conservative and towards the end of life imposes a further strength reduction of 30 – 40 % on the keyways. Without further investigation this method was considered to be the most appropriate at the time of calculation.

The final calculated capacities for both the keys and keyways are compared and the lower bound of the two is used for post processing. Currently, the post processing method assesses the contact loads against the capacity and every reported time step over the course of the assessment. A key/keyway contact is considered to have failed if the capacity has been exceeded during that instant.

CONCLUSIONS

This paper has provided details on the modelling approach used in creating a non-linear seismic model of the AGR reactors and the methodology of incorporating graphite shrinkage and irradiation effects. Only a number of interaction types have been considered in this paper, however, the modelling assumptions and data interpretation remain similar for other interactions within the seismic model, *e.g.* axial keys, vertical brick contact, *etc*.. In summary:

- Only a small proportion of the available material data output from the Serco brick analysis is incorporated into the seismic model.

- The global seismic model incorporates a number of axial and radial layer groups to capture the effects of decreasing dose outwards from the inner core. Form factors are used to calculate an equivalent dose and the data interpolated from the peak rated Serco brick model.

- Radial brick shrinkages from three data nodes are averaged across the height of the bricks and used to calculate the contact clearance between horizontally adjacent bricks. Local differences in radial shrinkage along the brick axis are not captured explicitly in the seismic modelling.

- Incorporation of keyway pinching in the seismic model uses a single data point at the position of the midpoint of the key. If an average keyway shrinkage was used the effects and timings of keyway pinching would be overestimated.

- Irradiation effects on the keys and keyways are averaged across the height of the brick and applied to the physical test capacity. In addition, due to the varying effects of fractional remnant strength on the keyways only strength loss over the key engagement length is incorporated.

- The increase in key/keyway prying forces due to changing brick geometries is simplified to a three point contact situation. Although considered conservative no further work has been undertaken to gain a better understanding of the dynamic effects of this phenomenon.

Acknowledgements
The authors would like to thank EDF Energy for financial support and permission to present this paper. The views expressed in this paper are those of the authors and do not necessarily represent those of EDF Energy.

References

McLachlan, N. (2003). Compendium of CAGR core and sleeve data. British Energy Document CSMDC/P28 Data sheet C2/2

Slater, I. G. (Oct 2009). AGR graphite UMAT – ABAQUS user material, subroutine for isotropic graphite. Serco report, SA/SIA/20022/R003 Issue 3

Solid-Body Simulation of the Dynamic Response of an Array of Graphite Bricks

Steve Brasier and Stuart Rogers
Atkins, 500 Park Avenue, Aztec West, Almondsbury, Bristol, BS32 4SY
Email: steve.brasier@atkinsglobal.com

Abstract
Modern commercial Finite Element Analysis codes are now incorporating contact simulation capabilities of increasing sophistication, but solid models of Advanced Gas-cooled Reactor cores containing tens of thousands of components are still too computationally expensive to be practical for seismic analysis. The performance of the experimental SOLFEC code under development at Glasgow University may make such problems tractable, but its use for AGR seismic assessments will represent a significant advance in the state-of-the-art. As an intermediate target, this paper describes the steps required to analyse the response of a smaller array of 81 core bricks which has previously undergone physical testing. Particular emphasis is given to the data and measurements required to support such analyses.

Keywords
Seismic analysis, contact simulation, SOLFEC

INTRODUCTION

The cores of the UK's Advanced Gas-cooled Reactors consist of tens of thousands of graphite bricks arranged in a multi-layered grid. These bricks provide moderation of neutrons to sustain nuclear fission and direct the cooling gas which carries heat from the fuel to the boilers. The bricks are connected in the horizontal plane by a keying system using loose or integral keys with corresponding keyways in the flanks of the bricks. Vertically-adjacent bricks also interact through a set of interlocking features on their end faces. At the start of life small clearances exist between bricks and in the keying system, but over the lifetime of the station irradiation-induced dimensional change is expected to reduce some clearances and increase others significantly.

The response of these cores to seismic motions is currently predicted using time-history dynamic simulations of 'stick and spring' models. In these models, bricks are represented by rigid bodies with assigned mass properties. The interactions between bricks (both through direct contact and via the keying system and end-face features) are modelled using parallel spring/damper pairs. These springs and dampers are assigned non-linear properties which include the 'lost motion' due to the clearances in the system.

The main advantage of this method is that it reduces the problem to a size which is actually solvable in a reasonable amount of time for a 10+ second seismic event. However with larger clearances due to irradiation-induced shrinkage or damage to core components, it becomes increasingly difficult to ensure all possible interactions within the core are captured by such models.

A solid modelling approach including contact is therefore attractive. In such an approach the full 3D geometry of all core components is simulated and contact interactions are detected and resolved automatically. Potentially any and all interactions arising between components can therefore be simulated, including those between damaged components. However, AGR core simulations represent an exceptionally large contact problem due to both the number of bodies and length of event to be simulated. Solid contact models in current commercial codes do not appear to scale well to this size of problem. Even if massive parallelisation could deliver the required computing performance, per-CPU core licensing makes this prohibitively expensive.

The experimental SOLFEC code under development at Glasgow University implements a novel form of contact constraint using the implicit Contact Dynamics method as described by Koziara (2010). It is specifically oriented towards solving dynamic frictional contact problems using large-scale parallel processing, raising its potential to provide solutions to this class of problem in a practical timeframe. In addition it has no licensing costs for use. However, the current immaturity of the code, the use of a novel contact constraint approach and the small user base at present mean that its use as a routine engineering tool for AGR core seismic assessments lies some way off.

As an intermediate step towards this goal some existing physical tests have been simulated using SOLFEC. These tests used a single layer array of 81 core bricks mounted on a shaker table to investigate the response of the array. This paper describes how such simulations may be performed, with particular emphasis on the data and measurements required to obtain accurate results. It also discusses additional aspects which would need to be considered for actual AGR core seismic assessments.

DESCRIPTION OF TESTS

A series of dynamic tests were carried out by National Nuclear Corporation in 1985 as part of the seismic endorsement of the Heysham II / Torness AGR core design. These tests used an array of 81 graphite bricks as shown below, consisting of alternating 'fuel' and 'interstitial' bricks. The array was mounted in a rigid frame on a shaker table.

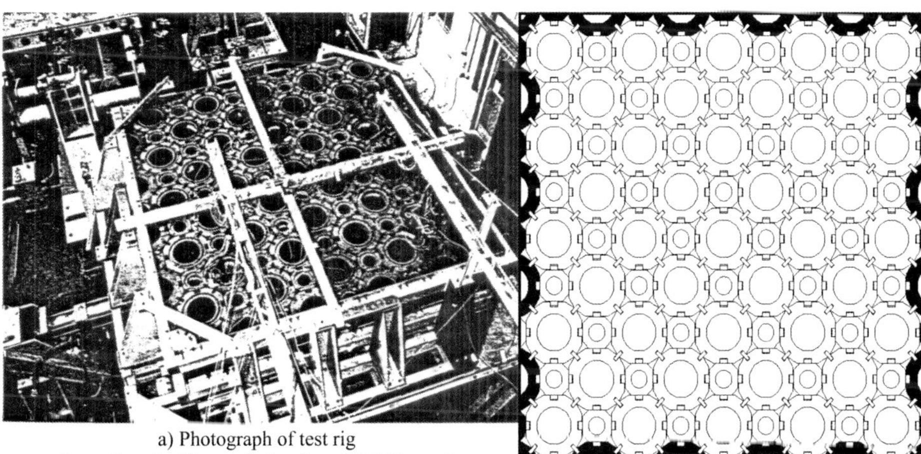

a) Photograph of test rig
(reproduced with permission from EDF Energy)

b) Diagram of array; motion applied via shaded partial bricks on boundary

FIGURE 1: 81-brick test arrangement.

The boundary frame was driven in the horizontal plane with synthesised seismic or swept-sine input motions in either one or both axes simultaneously. Measurements were taken of the resulting velocities of certain bricks in the array (determined from accelerometer data) and also of the forces within the keying system (using load cells built into some components). An example of the type of results produced is reproduced below from the test report.

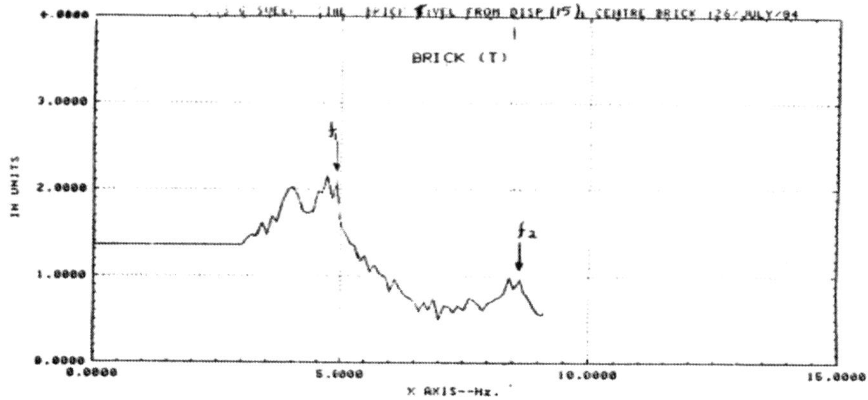

FIGURE 2: Example of test results: velocity of central brick in array against input frequency for a swept-sine input (reproduced with permission from EDF Energy).

Ideally, a simulation of these tests would be able to reproduce the response of the array to the input in terms of both velocities (as shown above) and forces. As an initial stage, only the velocity response has been considered. For actual AGR core seismic assessments the response in terms of brick displacements will also be of interest.

PROBLEM PHYSICS AND SIMULATION

The simulation of the above problem may be broken into three main areas of interest here:

- Geometry.
- Mass properties.
- Contact detection and enforcement (contact interactions).

Other aspects such as the input motions and boundary conditions to be applied are generally straightforward and in line with conventional Finite Element Analysis practice.

Simulating the geometry of the bricks and keys in the tests does not present any significant difficulties, although some practical problems were encountered with the test documentation. The only specific modelling issue is that curved surfaces are represented with faceted surfaces (as for standard Finite Element methods). This can lead to issues if contact is expected to occur on these surfaces, as the position of the contact could potentially be different between the real and modelled situations. Actual AGR core components will have a somewhat more complex geometry, which may lead to a more refined mesh and lower computational performance. The real difficulty for actual AGR core seismic assessments is probably the uncertainty in the geometry to be modelled, which will arise from uncertainties in the original 'as-built' geometry (e.g. tolerances), irradiation-driven dimensional changes and thermal expansion.

The mass properties of the bricks and keys are clearly important in determining the dynamic response of the array. For a 3D solid modelling approach such as SOLFEC the appropriate mass properties are computed automatically using the geometry, and only a material density needs to be specified. For this work density was back-calculated from the reported component masses. Although this resulted in slightly different densities for the different types of brick and loose key there are not expected to be significant issues with the accuracy of either measured or computed mass properties for these tests. For actual AGR core analyses the situation will be significantly more complex due to oxidation of the graphite during irradiation, leading to non-uniform density within components and significant variations in component masses. Some means of predicting the density distributions and the related uncertainties and variabilities will therefore be required.

The contact interactions are the novel part of this analysis and hence merit more detailed discussion. This is provided by the next section.

CONTACT INTERACTIONS

The contact interaction between two bodies in the array may be broken down into normal and tangential components. The tangential component is due to friction. While simulating friction can be a mathematically harder problem than the normal contact behaviour these difficulties are generally hidden from the analyst. SOLFEC implements a Coulomb friction model and modelling therefore requires:

- Demonstration that a Coulomb friction model is appropriate for this problem – for this work this was assumed in the absence of contrary evidence.

- Specification of an appropriate friction coefficient. Available test data shows large variability in graphite-graphite friction, even for known graphite grades in an un-irradiated state. However preliminary work suggested that as long as the model was not frictionless, displacements were unaffected by the value of the coefficient and velocities were relatively insensitive. Therefore a value of 0.1 was used which lay near the centre of the range of available data.

The tangential component of the contact interaction is not discussed further here. The normal component of the interaction has two main aspects to consider; *interpenetration* and the *restitution coefficient*. These are discussed in turn below.

Interpenetration

Interpenetration, the extent to which bodies overlap during the simulation, occurs with various approaches to contact simulation. It does not have a physical basis and hence does not require any experimental data. SOLFEC determines where contact constraints should be applied to the model at each time step by detecting the overlapping volumes between pairs of bodies, so some degree of interpenetration is necessary. However unlike some other contact algorithms this interpenetration should not grow once contact has been detected. SOLFEC also has an advantage over the commercial Abaqus / Explicit code in that the interpenetrations between bodies can be reported as an output variable of the simulation, as shown overleaf.

What constitutes an acceptable level of interpenetration is not entirely obvious. It would seem appropriate that it is small relative to the maximum gap between surfaces which come into contact, but for some pairs of surfaces with some array geometries this gap may itself be very small. It is expected that displacement results would be sensitive to the interpenetration, but for the 81-brick tests no displacement data is available. Preliminary work suggested that velocity results were not particularly sensitive to the level of interpenetration, as long as the overall behaviour of the bodies appeared physically reasonable, *e.g.* keys did not pass through the sides of their keyways. This was affected by the analysis time-step specified and so the approach to controlling penetrations was simply to specify a small enough time-step that the modelled array behaviour appeared physically plausible on visual inspection.

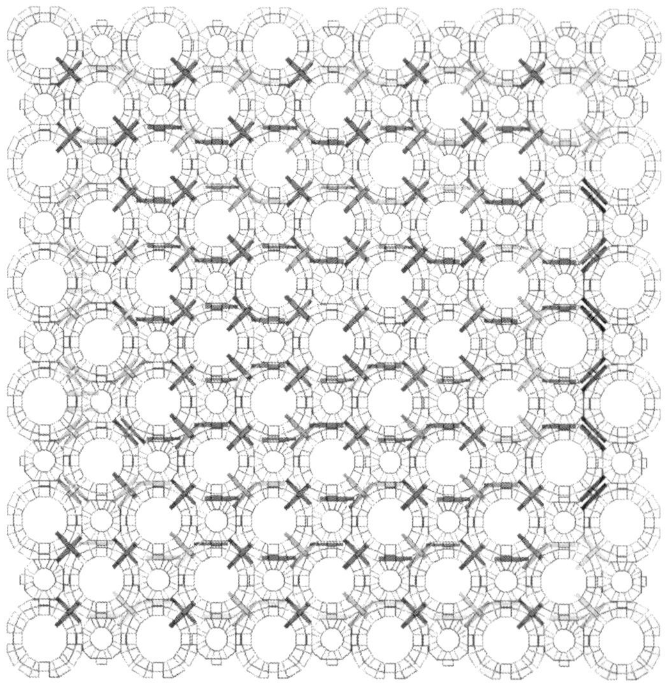

FIGURE 3: Example of interpenetrations: shaded bars indicate overlapping volumes (0 – ~0.16mm).

Restitution

The coefficient of restitution (e) of a collision is the ratio between the relative velocities of bodies before and after the collision. It lies in the range 0 to 1:

e* = *1: An elastic collision in which kinetic energy is conserved.
e* = *0: A perfectly inelastic collision. Kinetic energy is not conserved and the bodies are stationary relative to each other after the collision.

In reality, all collisions will be inelastic with e lying between these extremes. As bodies come into contact they deform, converting kinetic energy into strain energy. Once they have been brought to rest relative to each other this stored strain energy drives them to rebound, converting strain back to kinetic energy. During this deformation, some of the strain energy will be dissipated as heat due to losses within the material, leading to a reduction in kinetic energy after the collision and hence an inelastic collision. Additional losses may occur at the contact interface due to friction. The deformation during the collision will also excite vibrational modes of the bodies which may continue after the bodies have separated, until these modes are damped out by losses within the material.

For the 81-brick tests, the restitution coefficients of collisions between the different combinations of bricks and keys was determined experimentally by measuring velocities in collisions between individual bricks constrained using linear bearings. These preliminary tests covered both direct collisions between the bricks themselves and 'shearing'-type interactions through the keying system, for both the fuel and interstitial bricks. The testing of multiple collision configurations is important as each configuration would be expected to have different behaviour. A range of collision speeds from 0-30 cm/s was also covered, as the losses (and hence restitution) will depend on the energy in the collision. In all, 10 test configurations were used with of the order of 40 tests in some configurations – hence these preliminary tests represent a significant experimental programme in themselves.

For contact simulations, the restitution coefficient achieved arises from the behaviour of both the contact constraint itself, and the bodies in the contact. Penalty contact methods which generate a force as some function of the penetration (effectively inserting a stiff spring at the contact position) can provide non-zero coefficients of restitution even without any straining of the bodies themselves, and may therefore model elastic or inelastic collisions between rigid bodies. By contrast, SOLFEC's contact method will by default result in zero relative velocity at the contact interface. Collisions between rigid bodies will therefore have a restitution coefficient of zero, *i.e.* perfectly inelastic. Deformable bodies are required for collisions with non-zero restitution coefficients, with the rebound being driven by the deformation of the body and conversion from kinetic to strain energy and back again[3]. Stiffness-proportional damping is then applied to each body, resulting in losses in strain energy during deformation and hence an inelastic collision with a restitution coefficient < 1.

The restitution coefficient in a SOLFEC simulation is therefore affected by several properties of the simulation. These are discussed individually below but in general it is not feasible to state from first principals what these properties should be for a given restitution coefficient. Instead, the simulation must be tuned until the desired restitution coefficient is achieved. For this work, the tuning is carried out using analyses of individual brick collisions, similar to the preliminary tests used to measure the experimental restitution. Some combinations of properties may result in an unstable solution. As SOLFEC does not check stability automatically so the analyst must try various combinations of parameters and develop an understanding of where the stable region lies for a particular analysis.

Time step:
The size of the time step affects which vibrational modes of a body are represented in the simulation. Larger time steps mean that higher modes are not captured, reducing the total energy and lowering the restitution. However smaller time steps increase the computational cost of the analysis. For these analyses the largest time step giving physically-reasonable

[3] In fact, a non-zero restitution coefficient for the actual contact constraints may be directly specified in SOLFEC. This feature of the code is experimental at present. With the Contact Dynamics method SOLFEC implements, restitution coefficients other than zero or one are not necessary energy consistent, and may result in energy being added to the solution which would be undesirable. With deformable bodies, such a non-zero contact restitution coefficient would also interact with any strain-driven rebound to determine the overall restitution of collisions. Given these issues this feature was not used for the work described here, although future work may consider it.

results with respect to the configuration of the array was therefore used, as discussed above under Interpenetration.

Body Type:
Individual bodies in a SOLFEC analysis may use rigid, pseudo-rigid or Finite Element Method (FEM) formulations. Rigid bodies cannot deform and are therefore unlikely to be suited to this problem as discussed above. FEM bodies use a conventional element formulation. The pseudo-rigid formulation is deformable, but has constant strain gradients across a body. The deformed shape is therefore a linear transformation of the undeformed shape. This results in far fewer degrees of freedom than a Finite Element body, and therefore much higher computational performance. SOLFEC also offers a rough background mesh formulation in which a coarse FEM mesh is used to calculate approximate body strains which are then interpolated onto a separate accurate geometry used for contact detection. This approach was not tested here due to the additional complexity but is worth further investigation. Preliminary work showed that Finite Element bodies were too costly for simulations of the full 81-brick array on the available hardware[4], hence the pseudo-rigid formulation was used. The choice of body type has further implications as discussed below.

Damping:
A stiffness-proportional damping coefficient may be specified for each body. This damping controls the energy dissipated during straining of the body, and hence directly affects the restitution. It is therefore the primary parameter used to tune the restitution coefficient to the desired values. Very low levels of damping may however cause stability issues, and may also result in the excited vibrational modes of the body persisting until the next collision, adding additional energy into this second collision.

Mesh:
For a FEM body the mesh defines the mass discretisation of the body and therefore affects the amount of mass associated with a contact constraint, and hence the kinetic energy lost as the surfaces are brought to relative rest. The mesh also defines the body stiffness discretisation which (due to the stiffness-proportional damping) will affect the strain energy dissipated during deformation of the body. In contrast for a pseudo-rigid body the body geometry may be defined by a mesh, but the mass and stiffness operators are computed directly from the body shape. Hence, they are independent of the mesh discretisation which therefore has no effect on restitution for a pseudo-rigid body.

Material modulus:
The Young's modulus of the material is an input parameter to the simulation if deformable (pseudo-rigid or FEM) bodies are used. For conventional static Finite Element stress analyses, the modulus used in the simulation is intended to correspond to the actual material modulus. The analyst then ensures the mesh used results in the correct body stiffness. Considerable effort has been expended by others on developing an understanding of both the virgin (un-irradiated) modulus of AGR graphite and the changes in modulus with irradiation, supported by an extensive test programme on irradiated material taken from test reactors and trepanned from AGR cores.

For these analyses the approach may be somewhat different, as only the body motion is of interest, not the local stresses. With FEM bodies the mesh discretisation may be fixed by the

---

[4] Dell DL580, 4x Xeon E7340 processors (16 cores in total), 16 GB memory.

need to achieve satisfactory analysis performance. For pseudo-rigid bodies the body stiffness operator does not depend on mesh discretisation, as described above, but the body may be extremely over-stiff compared to an equivalent FEM body using the same modulus. With either body formulation there is therefore a case for treating the modulus as a tuneable parameter, as long as reasonable deflections are still achieved. Even if modulus is not tuned, a precise value is not required[5]. Very little information was available on the grade of graphite used for these tests so the analyses used a modulus for unirradiated AGR graphite.

Probably of more use than detailed material modulus measurements would be accurate measurements of the force/deflection behaviour of bricks and keys under static loads. The results from such tests may then be compared with simulations to see the combined effect of material modulus, body formulation and mesh discretisation (if applicable). Such tests were in fact carried out as part of the work-up to the 81-brick tests, and have been used during the restitution tuning process to understand what aspect of the simulation is causing differences.

RESULTS & CONCLUSIONS

Both the tuning analyses for the restitution coefficients and the analyses of the full 81-brick array are underway at the time of writing. However, some preliminary results can be outlined:

- FEM bodies appear to be too computationally expensive at present. Whilst more powerful hardware could be used, the desire to simulate full AGR cores having ~10000 bricks (plus corresponding keys) suggests that models of an 81-brick array must be correspondingly computationally cheap, and therefore that pseudo-rigid or rough background mesh formulations are more appropriate at present. However, the performance of the FEM mode is still under active development so this approach may become more feasible in the future.

- The tuning analyses completed to-date suggest that using pseudo-rigid bodies tends to result in restitution coefficients towards the lower end of the desired range, when using the largest feasible time step (for computational performance) and appropriate levels of damping. Some changes to the analyses to mitigate this are being considered, and the effect that lower restitution coefficients have on the response of the 81-brick array will be shown by ongoing work.

- The 81-brick array analyses completed to-date show velocity responses of the same form as those from the physical tests, *i.e.* a peak response around 4 Hz dropping off to a much lower response at higher frequencies. This provides some confidence that solid modelling using SOLFEC is a feasible approach to simulating these types of problems.

[5] An AGR core seismic analysis would probably require sensitivity studies of this and other parameters as part of a quality assurance process, but such discussions are beyond the scope of this paper.

Acknowledgements
The authors would like to thank EDF Energy for funding the work reported here and permission to present this paper, and Tomaz Koziara and Nenad Bicanic at Glasgow University for their help and support with the SOLFEC code. The views expressed in this paper are those of the author and do not necessarily represent those of EDF Energy.

References

Ahmed, K. (1987). The non-linear seismic response of AGR core graphite brick slices – correlation of experimental and analytical results. *Earthquake Engineering and Structural Dynamics*, 15, 159-188.

Koziara, T and Bícaníc, N (2010). A distributed memory parallel multibody Contact Dynamics code. Retrieved from http://wiki.solfec.googlecode.com/hg/2010-Koziara-distributed-contact-dynamics.pdf

Intelligent Graphite Core Condition Monitoring

Christopher J. Wallace, Graeme M. West, Stephen D. J. McArthur and Michael Coghlan*

Insitute for Energy and Environment, University of Strathclyde, Glasgow

*GCPT, EDF Energy, Barnwood, Gloucester

Email: cwallace@eee.strath.ac.uk

Abstract

Condition monitoring of graphite cores has increased significantly in recent years, as the safety cases required to continue operation of Advanced Gas-cooled Reactors (AGR) have evolved. This paper describes the development and implementation of intelligent systems designed to automate and formalise the condition monitoring (CM) analyses and tasks performed within the remit of the Monitoring Assessment Panels, which are currently being rolled out across the AGR fleet. The implementation of such systems is found to depend crucially on the involvement of station staff at each step of design and implementation, which is highlighted by the successful deployment of second iterations of the IMAPS system for managing reactor observations and the BETA system for automatically analysing refuelling data. Though each system was initially based on an individual analysis or tasks, this paper describes more recent work to allow closer integration of the systems as they are developed, in order to maximise the use of the available data while minimising duplication of effort and required operator time.

Keywords

Monitoring, analysis, software

INTRODUCTION

The introduction of a formal monitoring regime in the Advanced Gas-cooled Reactor (AGR) fleet was designed to complement increasing inspection as the stations meet, and in some cases exceed, their original design lives, but continue to operate through lifetime extension. With monitoring, however, came a large increase in the volume of data that required analysis and storage such that station staff could make decisions based on the monitoring information derived from the core. To date, this problem has been tackled in three main strands, of varying maturity and scope, which will form the structure of this paper.

After a brief introduction to the Monitoring Assessment Panel (MAP) meetings, the deployment of a system called Intelligent Monitoring Assessment Panel System (IMAPS) is described. Now in a second iteration, IMAPS (Jahn, 2007) is used to store and manage Condition Monitoring (CM) observations across the AGR fleet. Emphasis is placed on the interaction with station and central engineering staff to improve the system between the first and second versions of IMAPS. An important component of monitoring regime is the analysis of refuelling data from the AGR, which is performed automatically by a system called BETA (West, 2006), also now in its second revision. Drawing parallels with IMAPS, it is shown that after a successful introduction and after feedback from staff, improvements and refinements were made to a second version of the software.

In the context of the monitoring regime, it is shown that IMAPS provides a framework for managing CM observations, while BETA provides special functionality for a particular analysis. The integration of analyses such as BETA within the IMAPS framework and the addition of further analyses is the topic of the final section of the paper which discusses recent prototyping work using a distributed monitoring system based on intelligent software agents and the use of multiple data sets to extract more detailed condition monitoring information. The paper ends with a discussion of future work, with an emphasis on integration.

MONITORING ASSESSMENT PANEL MEETINGS

The aim of the MAP meetings are to review all new monitoring data gathered during the MAP period for evidence which might suggest distortion of the graphite core. The MAP meetings consider a number of Class 1 parameters (thermal to neutron power ratios, fuel grab load trace analysis, control rod monitoring and fuel handling observations) as well as a number of Class 2 parameters including various alarms and moderator temperature. Introduced initially at the lead stations at Hinkley Point B and Hunterston B, the MAP is now being rolled out to other stations across the AGR fleet. As a station owned process, the documentation and analyses vary slightly between stations, however, all condition monitoring observations are graded (in order of increasing severity) as blue, green, amber or red.

INTELLIGENT MONITORING ASSESSMENT PANEL SYSTEM

The initial version of IMAPS was developed and prototyped initially as a standalone application, but after further development and discussion, was deployed as a web application accessible to all users on the EDF network with suitable permissions, as shown in Figure 1. Observations are entered by users with the appropriate expertise in a particular parameter and then validated by a different user in order to maintain a high QA grade on the data within the system. The tracking, management and re-grading of observations allows the system to present the user with a view of the data on a particular date, such that decisions made based on the data can be justified in the future.

With this basis for managing and grading observations established, the system was populated with live data and has been used at every MAP meeting since 2009. Based on feedback and discussions with EDF Energy staff who used the system, new features and refinements were proposed and discussed along with research focussed work that aims to make maximum use of the available data.

Development and Deployment of Version 2

Based on discussion with the users of the initial version of IMAPS, several suggestions and new features were discussed, with those which were agreed to most benefit the users prioritised for Version 2. As a regularly meeting panel, the MAP documentation varies in content, but not in structure between meetings, therefore a templating function was introduced to IMAPS. This facility allows the user to generate data sets, descriptive statistics and summaries of any sub-set of data contained within IMAPS. With a properly structured template document, the MAP forms can be automatically populated before the meeting, provided the data has been uploaded to IMAPS. This should greatly reduce the overhead of paperwork in preparation of MAP meetings and allow for a more straightforward rollout of the system to further stations. IMAPS now also allows the uploading of data from several file types as well as the previous method by which each observation was entered manually by a user. Each observation still requires a second user to validate the information, but using this 'bulk upload' feature, allows for significantly faster addition of data to IMAPS.

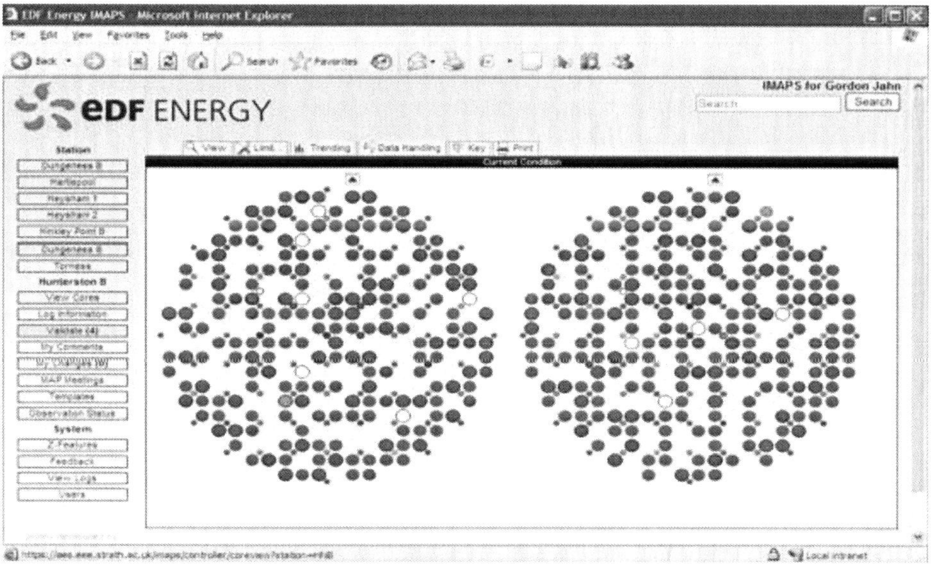

FIGURE 1: The IMAPS interface for a set of sample data.

As mentioned earlier, the MAP is a station owned process, therefore MAP documentation can vary between stations, however, there is significant commonality in the work undertaken by parameter specialists and the MAP chairman at each meeting. This observation led to the creation of a *roles* system, whereby an IMAPS user with sufficient permissions can create pre-defined user roles with a sub-set of the available permissions, such that users assigned a role can interact with IMAPS without providing unnecessary access to aspects of IMAPS not required by that user. This feature should allow for an easier adoption of IMAPS as the user base increases, due to the smaller fraction of the system that the user is required to be comfortable with before they use IMAPS on a regular basis. IMAPS Version 2 is currently being prepared for deployment by the IMD department at EDF Energy, and will imminently replace IMAPS Version 1 on the EDF network.

Trending and Clustering

As the volume of data stored within IMAPS increases, it was noted that there may be some merit in trending the occurrence of particular event types, grades or observations, therefore the ability to generate plots and statistics based on the data included in the IMAPS database, was included in Version 2. It was noted, however, that since the MAP panel grades only observations which were deemed to potentially indicate damage to the core, the total size of the data set within IMAPS is still relatively small, therefore the conclusions which could be drawn from such analyses are currently limited. It is expected, however, as the MAP process becomes more integrated across the AGR fleet, that useful trends may be detected and improve the understanding of the state and evolution of the core.

As a developmental feature not currently included in IMAPS, clustering was investigated as a method of analysis groups of observations. By comparing the location, time and type of particular observations, it is possible to group some observations into clusters, which can be explained by a single root cause. This method of analysis was selected due to the suspicion that it is possible that damage to the core may manifest itself in ways that are perhaps visible earlier in a statistical analysis than on the occasion of direct measurement of a parameter associated with the location of the damage such as during a refuelling event.

REFUELLING ANALYSIS

During refuelling of an AGR, the weight of the fuel assembly is measured as the old fuel is extracted and the new fuel inserted to ensure that the fuel moves smoothly through the fuel channel and is properly seated in the core. This measurement and the variation therein allow the inference of the friction acting upon the fuel as it moves through the internal surface of the fuel channel. The recorded measurement of weight is termed the Fuel Grab Load Trace (FGLT) and following every fuel movement, the FGLT is manually analysed and any potential indicators of core distortion reported formally through the MAP process.

To support the analysis of FGLT, the BETA system was developed which uses an encoded version of the engineer's understanding of refuelling behaviour to automatically interpret the FGLT for the presence of core distortion. The second iteration of the BETA software is now installed at EDF energy and is used to routinely assess refuelling undertaken at both Hunterston B and Hinkley Point B power stations.

Deployment of Version 2

At the time of writing, the BETA database now contains approximately 72,000,000 refueling measurements arising from 1540 fuel movements carried out at either Hunterston B or Hinkley Point B stations. BETA is being used to support the manual analysis of FGLT from both these stations, with expansion to Torness and Heysham 2 shortly.

Recent work on BETA has focussed on trending of refueling data, where trends in long term graphite shrinkage have been correlated in FGLT measurements (West, 2011).

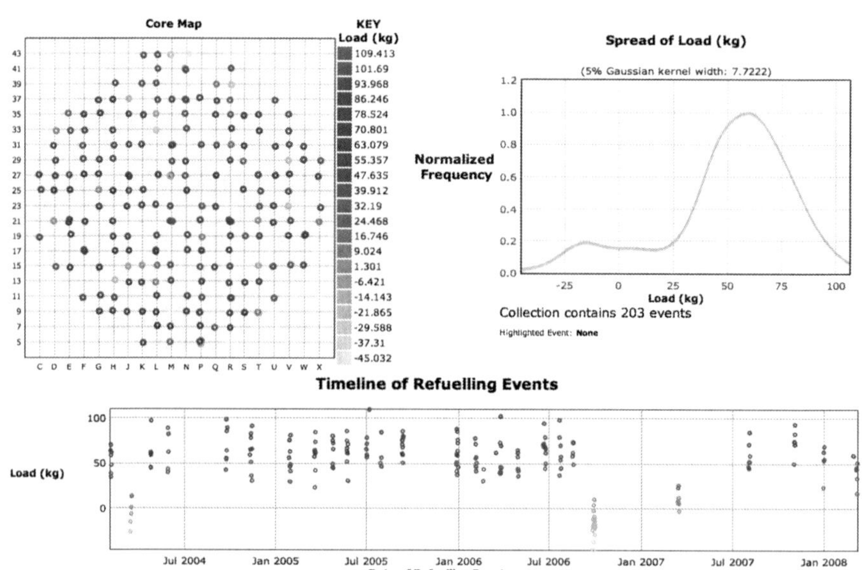

FIGURE 2: The BETA interface, showing the trending of loads across the core and with time.

Other work has focussed on the deconstruction of the FGLT traces into its constituent sources, allowing the frictional component from the lower stabilizing brushes to be isolated from the rest of the FGLT. This leads to a better representation of the underlying channel profile, permitting better detection of anomalies in the masked regions of the trace.

In addition, the data logger at Hunterston B has been upgraded to a new Yokogawa model. In order to manage this transition, it is important that the integrity of the data is maintained and that data captured using the new system can still be assessed and compared with the old traces. This was managed within BETA by implementing an additional data reading module at the front end. As a refuelling event is uploaded to BETA, the model of logger is detected and an appropriate conversion of the raw data is undertaken. This ensures that the data held within the BETA database is stored in common units to allow new and historical data to be compared.

Bringing uniformity and interoperability to the various CM analyses performed within the remit of the MAP has been an important aspect of recent work on BETA and IMAPS. In order to facilitate this work, investigations have begun into the practicability of implementing BETA within a more generic CM framework.

DISTRIBUTED AND INTEGRATED CONDITION MONITORING

A common feature of monitoring of older industrial equipment is the lack of a modern IT infrastructure, which when combined with the deliberate, nuclear safety guided limitations on networking and IT access of a nuclear power station, significantly complicates the collection, storage and dissemination of reactor data. In order to better resolve this problem, and provide an environment where new analyses or interfaces can be introduced without designing a new system from the ground up, recent work has investigated the viability of implementing a distributed CM system based on intelligent software agents.

Distributed Agent Framework

Intelligent agents are autonomous pieces of software which are programmed with particular expertise, such as '*how to analyse refuelling data*' or '*how to access and interpret the IMAPS database*', and can perform these tasks autonomously. With a community of these agents, which are programmed to be social and can cooperate to perform complex, distributed tasks more efficiently than a human, it is proposed that a distributed monitoring framework could be created. Such a framework would allow for faster and more repeatable condition monitoring analyses, while providing extensibility such that new analyses and data sources can be added as they become available with only minimal changes to the system.

Systems of many agents, or Multi Agent Systems (MAS) have been used previously in power systems and CM applications (McArthur, 2007), in a decentralised manner such that the system can be distributed over several physical locations or agent containers as nescessary, but remain federated through network access. Figure 3 shows a proposed structure for such a system with several containers of agents each with special expertise that they can share with the agent community. Agents can be encoded with domain knowledge that may include how to perform a particular analysis or how to access a particular data type, tasks which it can perform as proxy for any other agent in the system that makes a suitable request.

Ontology

In order for agents to communicate with each other, they require a common language or ontology. The ontology provides a structure to describe different classes of objects and domain knowledge in the system, such as reactors, control rods and FGLT data as well as allowing reasoning against these objects, for statements such as *a fuel channel is within a reactor* or *low power is less than x percent*. Figure 4 shows an example of what part of such an ontology looks like, showing examples of various physical components in a nuclear power station and how they are related. To demonstrate the suitability of CM analyses to deployment in a MAS, an example of an existing analysis performed as part of the MAP is demonstrated as an analysis agent within the prototype framework.

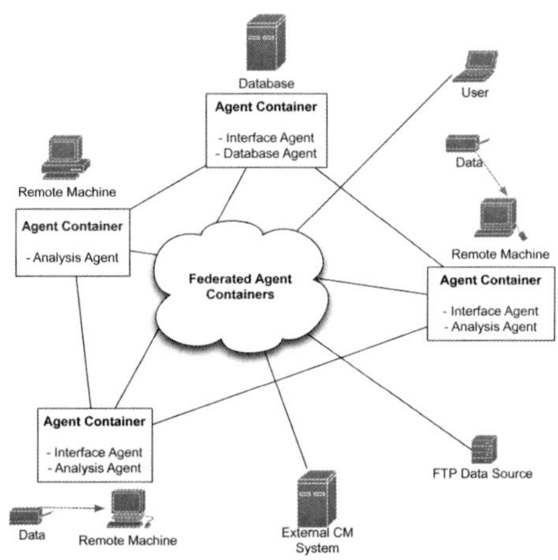

FIGURE 3: The proposed structure of a distributed Condition Monitoring system.

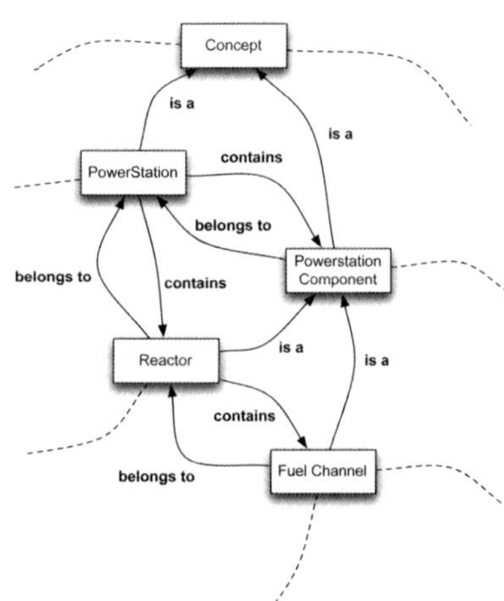

FIGURE 4: An example of a part of a nuclear condition monitoring ontology.

Thermal and Neutronic Power Analysis

Regular calculations are made of the channel powers for each AGR core, with a thermal value and a neutronic value calculated for each of approximately 300 channels every 7-10 days. Discrepancies between the two values are often due to modelling errors and core operations, such as refuelling or the exchange of control rods, however, an unexplained discrepancy could potentially indicate damaged fuel that is not being cooled properly. Each time sets of channel powers are generated, an engineer is required to compare the results with all relevant data sets and attempt to correlate events in order to explain the discrepancy. This is a time consuming task, but one that with sufficient data available, is algorithmically suitable to automation for a large percentage of anomalies.

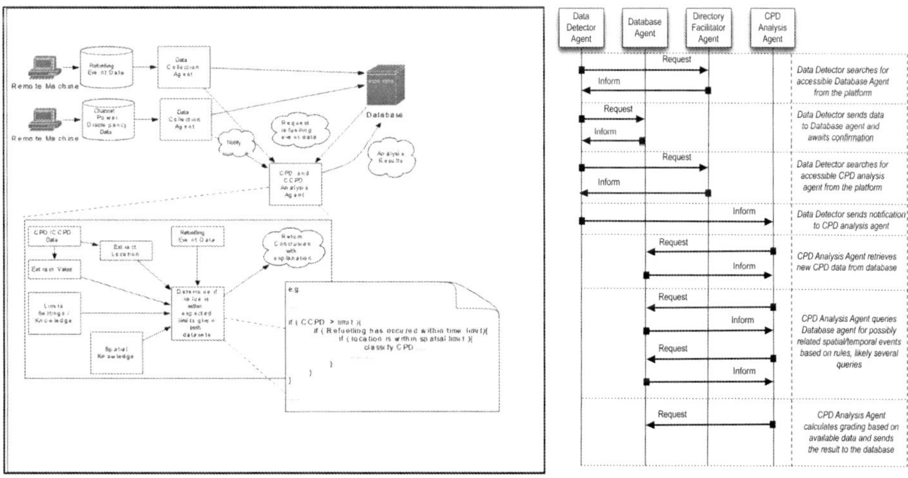

FIGURE 5: The process by which a set of agents can perform the channel power analysis, and on the right, the sequence of messages that are passed to perform the analysis.

Figure 5 shows how the analysis can be encapsulated in an agent, for the particular case of comparing refuelling records with anomalous channel power discrepancies. The analysis is performed as soon as the data becomes available, takes a fraction of a second and can be performed retrospectively on any amount of relevant data. The results of the analysis are then stored in a database in a form consistent with the ontology and can be used by other analyses or recalled or recalculated later. This example highlights the problem of having several large sets of data from disparate sources with varying availability that require correlation and matching of events, such that the interpretation of an event, in this case an anomalous channel power discrepancy, makes maximum use of available data as quickly as possible.

CONCLUSIONS

This paper has described the further development of the BETA and IMAPS systems for intelligent CM of the AGR fleet. It was shown that after successful introductions, both systems underwent further development based on feedback and close interaction with station and central engineering staff and a second iteration of each system is now ready to be deployed. The extension of both systems into further analysis of CM data as it is accumulated was shown to be the basis for future work on both systems. As the database of observations in the IMAPS database increases, the application of trending and clustering analysis to the data across the AGR fleet may yield new knowledge of the evolution of the cores, while the application of BETA to analysis of graphite shrinkage has already generated promising results.

Further analyses and closer integration for both systems has been proposed within a distributed CM framework, for which it suggested than a MAS approach is practical. A demonstration of the encapsulation of an existing MAP analyses within an agent has shown that it is at least possible to create such a framework, however further work is required to determine the practicability of such an installation within an AGR station.

Acknowledgements

The authors would like to thank EDF Energy for financial support and permission to present this paper. The views expressed in this paper are those of the authors and do not necessarily represent those of EDF Energy.

References

Jahn, J., McArthur, S. D. J., Reed, J. and Towle, D. (2007). Staged implementation of an agent based advanced gas-cooled reactor condition monitoring system. *Proceedings of IEEE Power Engineering Society General Meeting.*

West, G. M., Jahn, J., McArthur, S. D. J., McDonald, J. R. and Reed, J. (2006). Data mining reactor fuel grab load trace data to support nuclear core condition monitoring. *IEEE Transactions on Nuclear Science*, 53, [3].

West, G.M., McArthur, S.D.J. and Towle, D. (2011). Knowledge-directed characterization of nuclear power plant reactor core parameters. *Nuclear Engineering and Design*, 241, [9], 4013 - 4025.

McArthur, S. D. J., Davidson, E. M., Catterson, V. M., Dimeas, A. L., Hatziargyriou, N. D., Ponci, F., and Funabashi, T. (2007). Multi-agent systems for power engineering applications - part I: concepts, approaches and technical challenges. *IEEE Transactions on Power Systems*, 22, [2], 1743-1752.

McArthur, S. D. J., Davidson, E. M., Catterson, V. M., Dimeas, A. L., Hatziargyriou, N. D., Ponci, F., and Funabashi, T. (2007). Multi-agent systems for power engineering applications - part II: technologies, standards and tools for building multi-agent systems. *IEEE Transactions on Power Systems*, 22, [4], 1753-1759.

Decomposition of Refuelling Signals to Estimate Channel Bore Profiles

Graeme M. West, Christopher J. Wallace, Stephen D. J. McArthur and Michael Coghlan*

Insitute for Energy and Environment, University of Strathclyde, Glasgow

*GCPT, EDF Energy, Barnwood, Gloucester

Email: graeme.west@strath.ac.uk

Abstract

The fuel grab load trace (FGLT) generated during refuelling is a primary source of information relating to the structural health of the AGR cores. These refuelling events are routinely analysed and any anomalous behaviour reported at the station Monitoring Assessment Panel meetings. The FGLT is measure of the apparent weight of the fuel stringer as it travels through the graphite core. A number of components contribute to the measured weight, such as aerodynamic effects of coolant gas flow and friction caused by stabilising brushes on the fuel stringer interacting with the walls of the fuel channel. This paper describes a process for isolating each of these components from the FGLT. Separating out the effect of the lower stabilising brushes is particularly beneficial, as this provides a much-improved representation of the channel bore, and thus characterisation of any underlying defects. This is valuable for detecting anomalies in regions of the core where the net effect of the upper and lower stabilising brushes results in masking of the graphite brick bore profile.

Keywords

Condition monitoring, friction, fuel grab load trace

INTRODUCTION

This paper describes a model of the refuelling process consisting of three constituent components; a gas model, a model of lower stabilising brush friction and a model of upper stabilising brush friction. Modelled results are compared to both experimental results from a physical rig and from measured results from refuelling events at station and are found to be in agreement. This paper is split into the following sections. Firstly some background and related work are discussed before each component model is described in detail. The output from the combined model is compared to a measured response and the results are discussed. Finally, some proposed improvements to the model are detailed.

BACKGROUND

In-core inspections provide detailed information relating to the condition of the reactor core. However, inspections must take place during an outage, and therefore happen every two to three years, depending upon the station. It has been shown (West *et al.*, 2010) that Fuel Grab Load Trace (FGLT) data, gathered during routine refuelling, can provide additional information relating to the health of the core, and is considered a lead indicator of core distortion. Fuel grab load trace data for every refuelling event at Hunterston B and Hinkley Point B power stations is manually assessed for the presence of anomalies that may be indicative of core distortion. The results of this analysis fed into the formal quarterly Monitoring Assessment Panel (MAP) meetings held at the station, where the results are compared and discussed with other Class 1 parameters, such as thermal to neutronic power ratios, which also may indicate core distortion. The BETA system has been designed to provide automated decision support to the analysis of refuelling data. The BETA system compares the shape of a new refuelling event load trace against a benchmark of expected behaviour. This benchmark is derived empirically from previous refuelling events known to be free of anomalies. In addition, a knowledge-based expert system automatically identifies key features within the refuelling data and makes an assessment based on anomaly signature behaviour generated from experimental rig work of various crack configurations. West *et al.* (2006 & 2010) describe the BETA system in much greater detail.

In related work by Yang *et al.* (2010), a model-based estimation of the lumped friction effects and the aerodynamic effects from refuelling data is made. All the contributing friction is lumped together and modelled as one component and the remaining components were attributed as aerodynamic (gas) effects. The usefulness of the lumped friction model is limited as it does not effectively dissociate the effects of a single brush, and therefore the variation in bore, from the rest of the signal. However, the approach for approximating the aerodynamic effects provides a useful starting point for the model presented later in this paper. The contributions made in this paper examines multiple refuelling events and provides a generalisation of the channel-specific aerodynamic models detailed in Yang *et al.* (2010), allowing the generalised model to be applied to any refuelling event. In addition, the work presented here aims to isolate the contributions made by the lower stabilising brushes, thus provide a more accurate representation of the fuel channel bore.

Refuelling

The fuel in an AGR reactor is contained within a fuel stringer. Each fuel channel, created by an arrangement of graphite bricks that comprise the core, houses a single fuel stringer. During refuelling, the stringer containing spent fuel is removed from the reactor and exchanged with a stringer containing new fuel. During this fuel exchange, the weight of the fuel stringer is measured. The fuel stringer contains two sets of stabilising brushes, at the top and bottom of the stringer, which form an interface with the channel wall, resulting in a change in the measured load. The graphite bricks, which are stacked to form a channel, were straight at the commissioning of the station, but over time have distorted due to the effects of exposure to radiation. The level of distortion is key to determining the current health of the core and determining the internal bore measurements from the FGLT is a key driver behind this work. It is assumed that the reader has a working knowledge of both the refuelling process and the ageing process involved with AGR graphite, but more information can be found in Ng (1995) and Steer (2007), respectively.

Modelling the Refuelling Process

Three main components have been identified which contribute to the overall shape of the measured FGLT; the effects of coolant gas flow, the friction generated by the lower set of stabilising brushes interacting with the channel wall and the effects of the upper stabilising brushes interacting with parts of the refuelling channel above the reactor core, such as the guide tube. Modelling the contribution of each component is now covered in detail in the following sections.

Gas Flow Effects

In order to provide maximum availability of the station, the AGRs were designed so that they could be refuelled while the station was generating electricity, albeit at a reduced power load. Typically, this *low power refuelling* which occurs at some of the stations involves reducing the thermal output to around 40% of capacity and performing a fuel stringer exchange. The internal conditions of the core remain similar to those of full power operation in terms of pressures and gas flows, but with a reduction in the operating temperatures. In addition, the stations may be refuelled during a statutory outage, where the reactor is still pressurized with significant flows of coolant gas present. Finally, more recently, the reactor has been refuelled during an outage, but with the reactor in a depressurized state. This need has arisen from the increased requirement for inspections during the statutory outages. The flow of coolant gas is associated with the pressure and temperature of the core. The flow of coolant gas in and around the fuel stringer during refuelling contributes to the resultant FGLT and it is important to quantify and remove this effect from the overall FGLT if the frictional component is to be isolated.

Isolating the Gas Effects

Yang *et al.* (2010) surmised that for a given charge and discharge event for the same channel undertaken a short period apart and under the same reactor operating state, the difference between the two traces should be the result of the gas (or aerodynamic) effects, as the frictional components should be equal and opposite, and the weight of the stringer should be the same during charge and discharge. During refuelling, the forces acting on the fuel stringer include:

- mg, the weight of the fuel stringer due to gravity.
- F_{fr}, the lumped friction contributed by the upper and lower stabilising brushes.
- F_{aero}, the effect of the coolant gas flow.

The FGLT measurement, $F_{fglt\text{-}c}$, for fuel insertion, or charge event the measured force can be decomposed as:

$$F_{fglt-c} = mg + F_{aero} - F_{fr} \qquad [1]$$

and similarly for fuel removal, or discharge, the FGLT measurement, $F_{fglt\text{-}d}$, by

$$F_{fglt-d} = mg + F_{aero} + F_{fr} \qquad [2]$$

The difference between Equations [1] and [2] reflects the direction in which friction acts in each case; during charge, the friction supports some of the weight of the fuel, whilst during discharge the friction restricts the motion of the fuel. Summing the two FGLT measurements on a point by point basis, assuming that the weight of the fuel stringer, mg, is constant and the friction force is equal in magnitude, but opposite in direction, results in

$$F_{aero} = \frac{F_{fglt-c}+F_{fglt-c}}{2} - mg \qquad [3]$$

Figure 1 shows the plot of a charge and a discharge trace, along with the resultant F_{aero} for a low-power refuelling event. Figure 2 shows the plot of a charge and a discharge trace, along with the resultant F_{aero} for an offload, depressurized refuelling event. The resultant F_{aero} approximates to a straight line, indicating that the coolant gas flow is not greater enough to significantly affect the resulting FGLT. The assumption that the friction effects are equal and opposite are not quite true. For example, at the brick interfaces there is an offset static friction effect at the interface that is direction dependent. As a result, peaks appear in the resultant F_{aero} that are artefacts of the friction interaction instead. To create a general model for the gas effects, the F_{aero} force was generated and plotted for 200 low power and offload, pressurized events. Figure 3 shows a plot of these (with the data normalised). As can be seen from the diagram, the traces follow the same overall pattern, although they vary in magnitude.

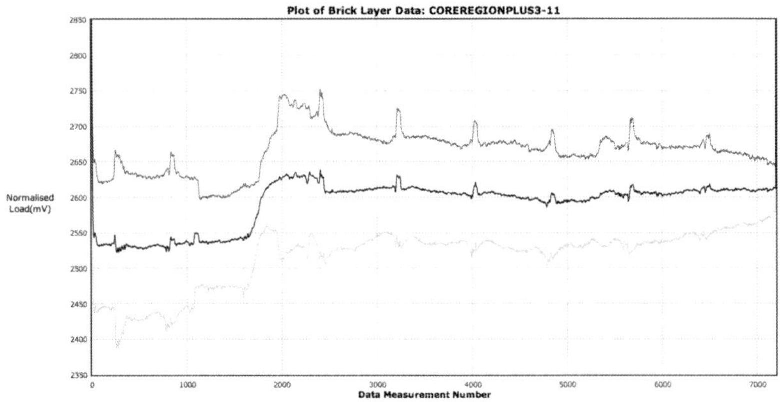

FIGURE 1: Plot of charge, discharge and resulting aero forces for a low-power refuelling event.

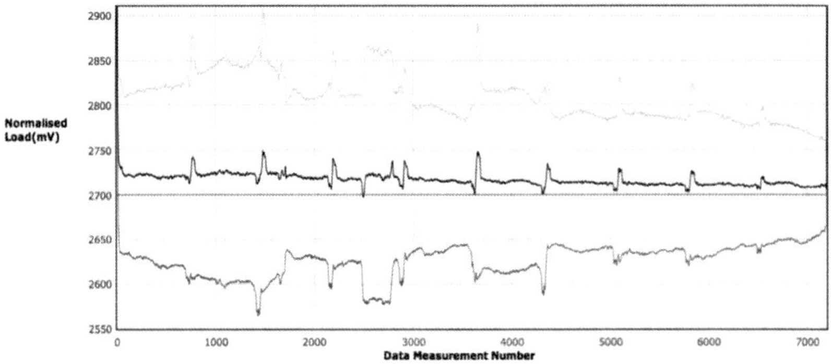

FIGURE 2: Plot of charge, discharge and resulting aero forces for an offload, depressurized refuelling event.

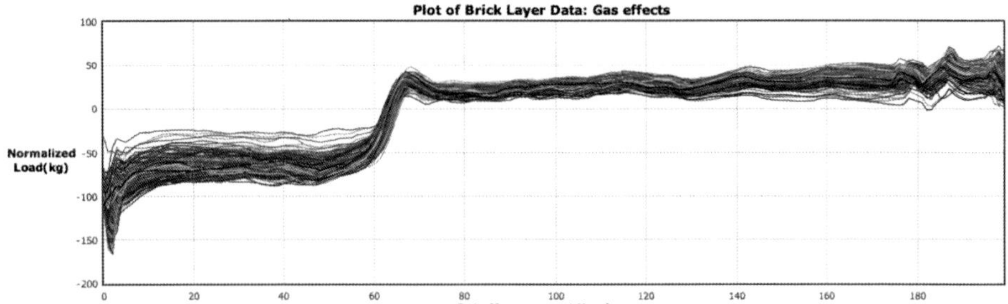

FIGURE 3: Plot of gas effects from 200 traces.

The average response for all these traces was calculated, and normalised between 0 and 1 to be fed into the overall model later, with a scaling factor to adjust the gas effect. The magnitude of this scaling factor will depend upon the reactor conditions.

The contribution made by the coolant gas, F_{gas}, is modelled using the following:

$$F_{gas}(x) = \begin{cases} 0, & 0 > x > H_1 \\ f(x), & H_1 > x > H_2 \\ 0.8887, & H_2 > x > H_3 \end{cases} \quad [4]$$

where f(x) is a function derived empirically from the data, and H_1, H_2 and H_3 are predefined heights in the channel.

Friction: Lower Stabilising Brush

A set of stabilising brushes on the nose of the fuel stringer, termed the lower stabilising brushes (LSB), guide the fuel stringer through the fuel channel. Figure 4 shows (a) a cross-sectional diagram of the brushes, (b) a photograph of a cross-section of the brushes and (c) a cross-sectional diagram of the fuel stringer nose, including the brushes.

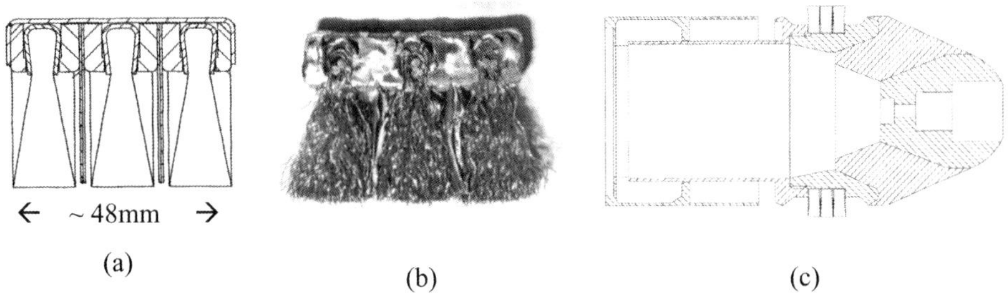

FIGURE 4: (a) Drawing of lower stabilising brush assembly; (b) photo of brush assembly; and, (c) fuel stringer nose indicating location of stabilizing brushes.

The diameter of the LSB is greater than the diameter of the fuel channel and as a result, the overlap creates a frictional component. This contains both a static component and kinetic component. Initially, considering the kinetic component in isolation, the magnitude of frictional force is proportional to the force normal to the direction of motion, and normally this is a component of the weight of the object being pulled along a surface. In the case of the fuel stringer traversing the core, the weight of the fuel stringer acts parallel to the direction of motion and therefore the weight of the stringer does not affect the frictional component. Instead, the normal force is generated by the restorative force of the brushes trying to straighten out while constrained by the smaller bore diameter of the channel. As the channel bore decreases the restorative force increases. Through experimentation the characteristic of brush friction, F_{br}, against channel bore, D_{bore}, at any given point, x ,is approximated by the 3rd order polynomial:

$$F_{br}(x) = \begin{cases} -0.025 \times D_{bore}(x)^3 + 18.826 \times D_{bore}(x)^2 - 4729.358 \times D_{bore}(x) + 396226, & D_{bore}(x) < 266mm \\ 0, & D_{bore}(x) \geq 266mm \end{cases} \quad [5]$$

A second assumption made for the purposes of simplifying the model is that where there is a gradual change in channel bore, such as the barrelling caused by the aging of the graphite, there is no associated static friction introduced. However, where there is a step change in diameter, such as at the brick interfaces, then static friction also needs to be considered. The exact boundary limit of what constitutes a step change is unknown, but for the purposes of this modelling step changes of over 1.5 mm are considered to introduce a static frictional component.

Experimental work by Skelton (2007) built a rig to simulate the friction response of the LSB. A surplus graphite brick from Dungeness B, was machined to span a range of fuel channel bore sizes from 251 mm to 263 mm in 2 mm steps and a test set of brushes was pulled through the brick to generate a load response. The purpose of this experiment was to determine whether bore shrinkage would lead to a large enough frictional component to cause the refuelling load reactor protection system to trip. The conclusions from the experimental work suggested that even brick shrinkage to very small diameters would not cause a reactor trip. The dotted line in Figure 5 shows the response obtained through one experiment. At each step change in diameter, the effect of each of the three brushes can be seen. Also shown in Figure 5 is the response obtained from a mathematical model of the experiment.

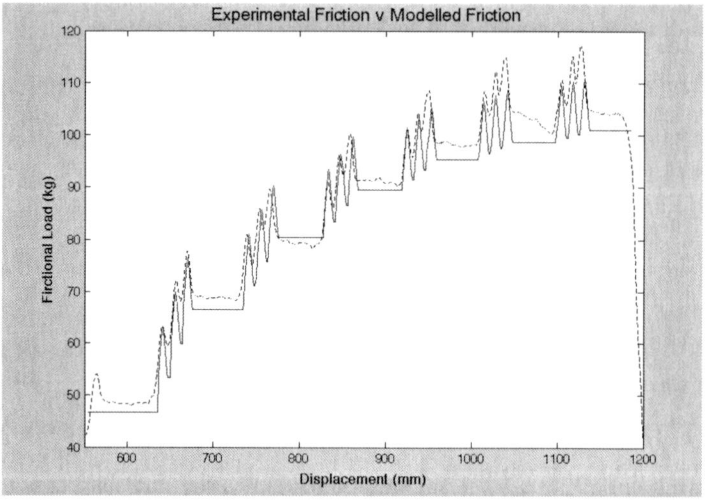

FIGURE 5: Friction obtained from an experimental rig (dotted line) and friction obtained from a simulated response in MATLAB (solid line).

The mathematical model was produced using the following steps:

1. The input to the model was an array of bore diameters stepping down in 2 mm steps from 263 mm to 251 mm, each step spanning 117.5 mm, emulating the dimensions of the machined brick used in the experiment. The resolution of the array was one data point every millimetre of displacement.

2. A second array, called the static impulse array, was generated which stores the location of any step changes above a certain limit by comparing consecutive points in the input array. In this case the limit was set to 1 mm. When a step change was found a roughly triangular feature with amplitude of 1, simulating the standard response obtained through static friction, was entered into the array. This represents the static friction component of the total friction and can be adjusted using a scaling parameter.

3. The kinetic response at any given point for each brush was generated from the experimentally derived characteristic equation given in Equation [5].

4. The overall friction was calculated for each value in the input array by summing the components of static and kinetic friction from all three brushes (the value of bore for each subsequent brush was displaced by 16 mm, the approximate width of a brush).

As can be seen from Figure 5, the modelled and experimental responses are in approximate agreement. Differences between the two responses are likely to be as a result of Equation [4] not being a perfect representation of the frictional response of the brushes. However, at this stage it provides confidence that the frictional response of the LSB has been reasonably modelled.

Friction: Upper Stabilising brush

When the fuel stringer is completely inserted into the core, the upper stabilising brush (USB) is situated towards the bottom of the steel guide tube, but not actually low enough to interact with the graphite core. The guide tube is constructed of several sections of steel pipe and finishes in a tundish, which exits into the "hotbox" region above the core. The lower section of the guide tube which connects to the top of the graphite core has a diameter of 248 mm, with the upper stabilising brush of ~250 mm, which results in this 1 mm radial overlap contributing a small frictional component. The guide tube then expands to a diameter of 260 mm, at which point there is clearance between the brush and the guide tube before entering the piston seal bore (PSB) section. The PSB section is also 248 mm and also contributes a frictional component to the load. This is followed by another 260 mm section that leads into the tundish and then into the "hotbox" region. For modelling purposes, it is assumed that a constant frictional load will appear in the FGLT in the regions corresponding to the USB in the guide tube and the USB passing through the PSB. To simplify the model, it is assumed that the USB will only contribute kinetic friction and that there will be no static component. Furthermore, it is assumed that the frictional load will be directly proportional to the difference between the USB brush diameter D_{USB} and bore diameter D_{bore} where the bore diameter is smaller than the USB brush diameter and zero when the bore diameter is greater than the brush diameter (no frictional interface). More formally, the friction force of the USB, F_{USB} at height x is given by:

$$F_{USB}(x) = \begin{cases} k * \left(D_{bore}(x) - D_{USB}(x)\right), & D_{bore}(x) > D_{USB}(x) \\ 0, & D_{bore}(x) \leq D_{USB}(x) \end{cases} \qquad [6]$$

Combined Model

The three model components, the model of gas effects, the LSB friction model and the USB friction model were combined to generate a modelled output for a given input based on channel bore measurements taken during inspection coupled with the choice of either pressurized or depressurised refuelling conditions. The results of one such refuelling are shown in Figure 6 with the modelled response (dotted line) compared to the actual measured FGLT response from an on-load refuelling event.

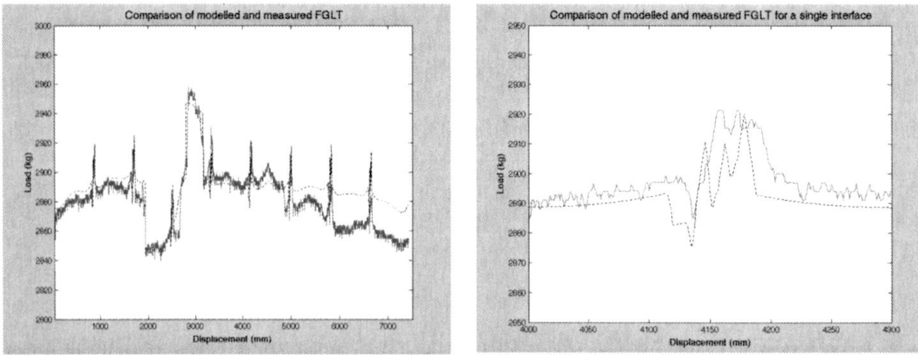

(a) (b)

FIGURE 6: Comparison of measured FGLT (solid line) with modelled FGLT (dotted line) for a (a) single discharge refuelling event over the full core; and, (b) zoomed to a single brick interface.

By visual inspection, there is a good agreement between the modelled response and the measured response, particularly for the lower brick layers towards the left of Figure 6(a). Over the last three brick layers shown in Figure 6(a), there appears to be a discrepancy, perhaps indicating there is an additional frictional component required. However, at this stage there is no explanation for this discrepancy and further investigation is required. At the interface between the brick layers, shown in Figure 6(b), there appears to be a slight offset between the model and the measured responses, as well as an additional small step. This is likely to be due to limitations in the frictional model and specifically the difficulty in modelling the large but brief diameter step change at the brick interfaces.

In addition to generating expected FGLT responses given the FGLT, it should be possible to generate the expected CBMU response from a given FGLT using this model. However, this will require Equation [4] to be rearranged to determine the bore from a given friction. This will allow permit unmasking of the FGLT to reveal the underlying shape of the channel bore. Wallace *et al.* (2011), has undertaken some preliminary work in this field.

Conclusions and Future Work

This paper has described a model of the refuelling process comprising three components: a gas model, a model of the lower stabilising brush friction and a model of the upper stabilising brush friction. The lower stabilising brush model has been compared to experimental rig measurements and provides adequate results, although improvements could be made to the model to better model the non-linear relationship between the brush stabilising brush deflection and the resulting friction. Reversing the mapping of the relationship between bore diameter and friction is the next step in the modelling process to allow the channel bore to be derived from measured FGLT data. A generalised model of the gas effects has been produced which allows both onload and offload refuelling events to be modelled. The model of the upper stabilising brush is simple but adequate for the current purposes, though static friction of the USB should be considered for completeness. Finally the model components have been combined to provide to provide an estimation of a FGLT from a given CBMU trace, and the modelled output agrees for the most part with the measured FGLT response, although an unexplained additional force, probably frictional, requires further investigation. In addition, though adequate, the model of the lower stabilising brush as it passes through a brick interface could be improved.

Acknowledgements

The authors would like to thank EDF Energy for financial support and permission to present this paper. In addition the authors would like to thank Jim Skelton from Amec for provision of the experimental rig data and associated photographs. The views expressed in this paper are those of the authors and do not necessarily represent those EDF Energy.

References

Ng, H. W. (1995). Dynamic analysis of AGR fuel assemblies. *Computers and Structures*, 54, [2], 339-350.

Skelton, J. (2007). Station core monitoring – experimental work to relate HPB/HNB grab trip load to bore shrinkage. Amec NNC Internal Report, Issue 4.

Steer, A. G. (2007). AGR core design, operation and safety functions. In:- Neighbour, G. B., (Ed.) *Management of Ageing Processes in Graphite Reactor Cores*, Special Publication No. 309, RSC Publishing, The Royal Society of Chemistry, Cambridge, UK, 11-18.

Wallace, C.J., West, G.M., Jahn, G.J. McArthur, S. D. J., and Towle, D. (2011). Data fusion for enhanced condition monitoring of nuclear reactor refueling. *Eighth International Conference on Condition Monitoring and Machinery Failure Prevention Technologies*.

West, G. M., Jahn, G. J., McArthur, S. D. J., McDonald, J. R. and Reed, J. (2006). Data mining reactor fuel grab load trace data to support nuclear core condition monitoring. *IEEE Transactions on Nuclear Science*, 53, [3], Part III, 1494-1504.

West, G. M., McArthur S. D. J and Towle, D. (2010). BETA: a system for automated intelligent analysis of fuel grab load trace data for graphite core condition monitoring. In: Neighbour, G.B. (Ed.), *Securing the Safe Performance of Graphite Reactor Cores*, Special Publication No. 328, RSC Publishing, The Royal Society of Chemistry, Cambridge, UK, 79-87

Yang, E., Grimble, M., West, G. M., Inzerillo, S., Katebi, R. and McArthur, S. D. J (2010). Model-based estimation and filtering for condition monitoring of AGR nuclear graphite cores. *UKACC International Conference on CONTROL 2010*, Coventry, UK. 1206-1211.

The LEWIS Method for Channel Functionality Assessment

Richard J. Crawford
Power and Process Europe, AMEC, Chelford Road, Knutsford, UK. WA16 8QZ.
Email: richard.crawford@amec.com

Abstract
A description is given of the LEWIS method of AGR-core channel functionality assessment. Using linear programming and the small-angle approximation, the method analyses the ability of control rods and fuel stringers to fit in a channel of the AGR core. The relative speed of this method enables assessment of every channel in a whole-core model, which enables the production of core maps and the ranking of channels. This, in turn, assists the basing of safety cases upon realistic models of core damage and loading, as opposed to previous bounding scenarios.

Keywords
Channel functionality, linear programming

ENGINEERING BACKGROUND

It is necessary to provide safety cases for the AGRs, which, among a range of topics, demonstrate the continuing safety of the graphite moderator cores over a future period against the fundamental nuclear safety requirements, *i.e.* unimpeded movement of both fuel and control rods and the provision of adequate cooling of the fuel and moderator, which are principally related to the geometry of the fuel and control rod channels. It is also desirable to be able to perform assessments of the economic lifetimes of the stations, in advance of rigorous safety cases, as maintaining functionality of the plant may be a more onerous requirement. To serve both requirements, it is necessary to have a method to ascertain the period within which the reactors will remain within their safety and functional criteria.

The AGR core contains a number of channels designed to hold either a fuel assembly or a control rod (both referred to as "assemblies" in this paper). The channels must allow the assemblies to be inserted and extracted as part of the operation of the reactor. Changes associated with the aging of the core can result in the channel shape distorting. This could potentially impair the ability to insert and extract assemblies.

The assessment of the core control rod and fuel stringer functionality has to date been performed using a sequence of bounding channel shapes at bounding amplitudes. The functionality assessment will no longer be restricted to a few bounding channel geometries, but will look at predicted channel geometries with a variety of channel distortions obtained from whole-core modelling and analysis. Therefore, there is the potential for a large number of load cases to be analysed. This analysis is very time consuming so there is the need for a filter to reduce the number of cases that must be analysed in detail. The method known as LEWIS has been developed to meet this need.

MATHEMATICAL BACKGROUND

Linear Programming

Linear programming is a form of optimisation. It is the process of minimising a linear objective function subject to a finite number of linear equality and inequality constraints as proposed by Karloff (1991). For example,

minimise	$z = x_1 - x_2,$	[1]
given	$x_1 + 3\, x_2 \leq 4,$	[2]
and	$2\, x_1 - x_2 \leq 0.$	[3]

Linear programming is normally concerned with the efficient use or allocation of limited resources to meet desired objectives (Gass, 2003). LEWIS uses linear programming for a different purpose.

General Form of a Linear-Programming Problem

The general form of a linear programming problem, as proposed by Karloff (1991) is:

minimise	$z = \mathbf{c} \cdot \mathbf{x},$	[4]
given	$\mathrm{LE}\, \mathbf{x} \leq \mathbf{le},$	[5]
	$\mathrm{GE}\, \mathbf{x} \geq \mathbf{ge},$	[6]
	$\mathrm{EQ}\, \mathbf{x} = \mathbf{eq}$	[7]
and	$\mathbf{x} \geq \mathbf{o},$	[8]

where **c**, **le**, **ge** and **eq** are given vectors, LE, GE and EQ are matrices and **o** is the zero vector of appropriate length.

Standard Form of a Linear-Programming Problem

It follows that the standard form of a linear programming problem, as proposed by Karloff (1991) and Press (2007), is:

minimise	$z = \mathbf{c} \cdot \mathbf{x},$	[9]
given	$\mathrm{A}\, \mathbf{x} = \mathbf{b}$	[10]
and	$\mathbf{x} \geq \mathbf{o},$	[11]

where **c**, and **b** are given vectors, A is a matrix and **o** is the zero vector of appropriate length. Problems can be converted from general to standard form by the use of slack variables (Press, 2007).

Linear Programming Solvers

Solvers are available that can solve linear programming problems. The majority of these require the problem to be posed in standard form. Given the matrix A and the vectors **b** and **c**, a solver will return a vector **xmin**, a value of the vector **x** satisfying Equations [10] and [11] and giving the minimum value of z in Equation [9].

A GEOMETRICAL PROBLEM

This paper will consider channel shapes in two dimensions. These can be generated from three-dimensional shapes by projecting onto a vertical plane. The LEWIS method is extendible to consider three-dimensional shapes in full, but that is not covered here.

FIGURE 1: Channel and assembly with *dsf* = 1 (left), *dsf* = 2 (centre) and *dsf* = 3 (right).

Figure 1 introduces the concept of distortion scale factor or *dsf*. Consider the theoretical channel and assembly shown on the left of Figure 1. If the channel distortion is scaled by a factor of 2, it will look as shown in the centre of Figure 1. If the channel distortion is scaled by a factor of 3, it will look as shown on the right of Figure 1. However, this scaling by *dsf* = 3 is not actually possible. Figure 1 shows the channel and assembly overlapping in the right-hand image. Therefore, the maximum distortion scale factor that can be applied is between 2 and 3. The LEWIS method calculates this maximum distortion scale factor.

The height of each brick is of the order of a metre and yet the clearance between the assembly and the channel bore is of the order of two centimetres. This fact enables the use of the small-angle approximation and the linearisation of the problem. The small-angle approximation of:

$$\sin\theta = \tan\theta = \theta \qquad [12]$$

And

$$\cos\theta = 1 \qquad [13]$$

is made. Therefore the geometry is as shown in Figure 2. The following vectors describe the geometry:

xb, the horizontal position of the end of each brick (xb_1, xb_2, ...)
yb, the vertical position of the end of each brick (yb_1, yb_2, ...)
rb, the bore radius of each brick (rb_1, rb_2, ...)
xa, the horizontal position of the end of each assembly segment (xa_1, xa_2, ...)
ya, the vertical position of the end of each assembly segment (ya_1, ya_2, ...)
ra, outer radius of each assembly segment (ra_1, ra_2, ...)

These vectors are all indexed from 1 onwards. Therefore:

- The end below a brick has the same number as the brick itself
- The end above a brick has a number that is greater than the brick by 1
- The end below an assembly segment has the same number as the segment itself
- The end above an assembly segment has a number that is greater than the segment by 1.

The geometrical problem is specified by specifying the vectors **xb**, **yb**, **rb**, **ya** and **ra**. A feasible value of **xa** is found as part of solving the problem. Implicit in describing the geometry in this way is the assumption that there is no end-face shear between bricks or assembly segments. The method is extendible to allow for either or both types of end-face shear but to-date this has been seen as unnecessary.

A LINEAR-PROGRAMMING PROBLEM

The LEWIS method constructs a linear-programming problem in general form. The vector of variables is:

$$\mathbf{x} = \begin{bmatrix} \mathbf{xa} \\ dsf \end{bmatrix}. \quad [14]$$

That is the final element of $\mathbf{x}$ is the distortion scale factor and the other elements are the horizontal position of the end of each assembly segment. This horizontal position is translated by a large constant cp so that constraint [8] does not interfere with the shape of the assembly.

The objective-specifying vector is:

$$\mathbf{c} = \begin{bmatrix} \mathbf{0} \\ -1 \end{bmatrix}. \quad [15]$$

Therefore, the optimal solution is one that maximises the distortion scale factor. This maximum distortion scale factor is known as $m1dsf$. The general-form problem is converted to standard form using slack variables and solved using an off-the-shelf solver. The constraints in the linear-programming problem relate to the geometry of the assembly in the channel as described below.

Constraint of the Assembly Ends Against Brick Bore Surfaces

The variable:

$$\rho = \frac{ya_i - yb_j}{yb_{j+1} - yb_j} \quad [16]$$

is the proportion of the distance up brick j that segment end i appears. Figure 2, shows that the constraint on the right-hand side of the channel can be expressed as:

$$xa_i - cp + \max(ra_{i-1}, ra_i) \leq (1-\rho)\, xb_j + \rho\, xb_{j+1} + rb_j. \quad [17]$$

Multiplying the horizontal positions of the channel by the distortion scale factor *dsf* gives:

$$xa_i - cp + \max(ra_{i-1}, ra_i) \leq ((1-\rho)\, xb_j + \rho\, xb_{j+1})\, dsf + rb_j. \quad [18]$$

Rearranging gives:

$$xa_i - ((1-\rho)\, xb_j + \rho\, xb_{j+1})\, dsf \leq cp + rb_j - \max(ra_{i-1}, ra_i). \quad [19]$$

This is expressed in matrix form by taking the right-hand side as an element of the vector **le** and the coefficients of xa_i and dsf on the left-hand side as a row of the matrix LE. Most of the row contains zeros but column i contains a 1 and the final column contains $(1-\rho)\, xb_j + \rho\, xb_{j+1}$.

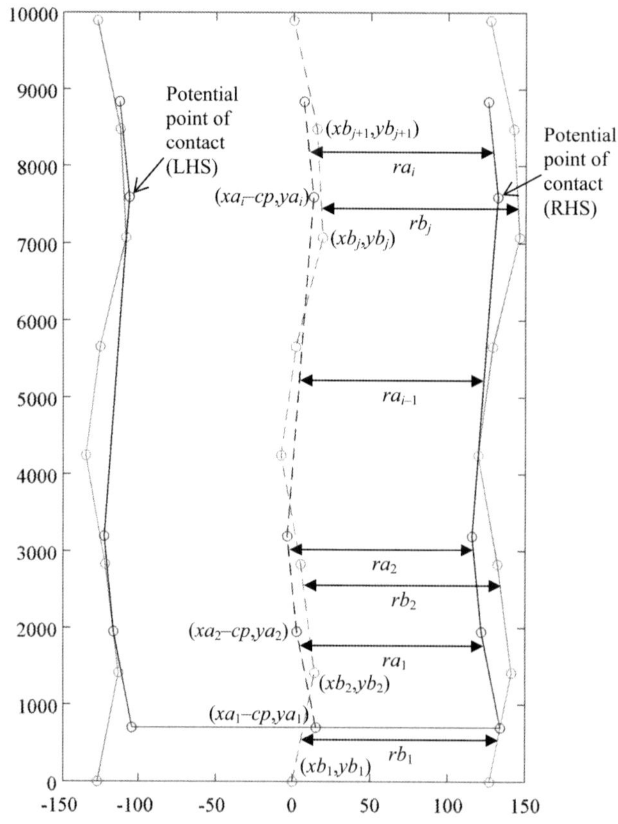

FIGURE 2: Constraint of the assembly ends against brick bore surfaces.

Similarly, the constraint on the left-hand side of the channel can be expressed as:

$$xa_i + ((1-\rho)\, xb_j + \rho\, xb_{j+1})\, dsf \geq cp - rb_j + \max(ra_{i-1}, ra_i). \quad [20]$$

This is expressed in matrix form by taking the right-hand side as an element of the vector **ge** and the coefficients of xa_i and dsf on the left-hand side as a row of the matrix GE.

Constraint of the Brick Ends Against Assembly Segment Outer Surfaces

This constraint is constructed in a similar way. For this constraint, the variable:

$$\rho = \frac{yb_j - ya_i}{ya_{i+1} - ya_i} \qquad [21]$$

is the proportion of the distance up segment i that brick end j appears. The constraint on the right-hand side can be expressed as:

$$(1-\rho)\, xa_i + \rho\, xa_{i+1} - xb_j\, dsf \leq cp + \min(rb_{j-1}, rb_j) - ra_i. \qquad [22]$$

The constraint on the left-hand side can be expressed as:

$$(1-\rho)\, xa_i + \rho\, xa_{i+1} - xb_j\, dsf \geq cp - \min(rb_{j-1}, rb_j) + ra_i. \qquad [23]$$

As with Equations [19] and [20], Equations [22] and [23] are expressed in matrix form using LE, **le**, GE and **ge**.

Constraining Angles of Articulation and Gapping

This paper will not go into details on constraining angles of articulation and gapping. Further inequalities can be used to constrain angles of articulation at individual joints. A constraint can also be put on the total gapping. However, that requires re-casting the problem in terms of angles of articulation.

ASSESSING CHANNEL FUNCTIONALITY

Assembly Height

LEWIS assesses the ability to insert and extract assemblies. Since LEWIS does not model forces, insertion and extraction are equivalent. By convention, LEWIS models withdrawal of the assembly from the channel. The distance from the bottom of the channel to the bottom of the assembly is recorded in the variable *assemblyHeight*. Figure 3 shows how the distortion scale factor depends on the position of the assembly within the channel in a typical case. The variable *m2dsf* records the minimum value of *m1dsf* found as the assembly is withdrawn from the channel.

Projection of 3D shapes

Of course, real channel shapes are in three dimensions. 2D LEWIS projects the three-dimensional shape onto a vertical plane at an angle of *psi* to the x-axis, as shown in Figure 4. Figure 5 shows a typical plot of *m2dsf* against *psi*. The minimum value of *m2dsf* over all values of *psi* is known as *m3dsf*. The value of *m3dsf* is used as a measure of channel functionality.

ASSESSING A WHOLE CORE

The relative speed of the LEWIS method allows the functionality of every channel in a whole-core model to be assessed. Figure 6 presents a scenario analysed by the AGRIGID whole-core modelling code. AGRIGID (Shaw *et al.*, 2008) calculates the displacements of the top and bottom of each brick in each channel. LEWIS assesses the functionality, giving an *m3dsf* value for each channel. These values are presented as a map of the core and two tables. The first table gives the worst 5% of channels and the second gives the percentiles at 5% intervals.

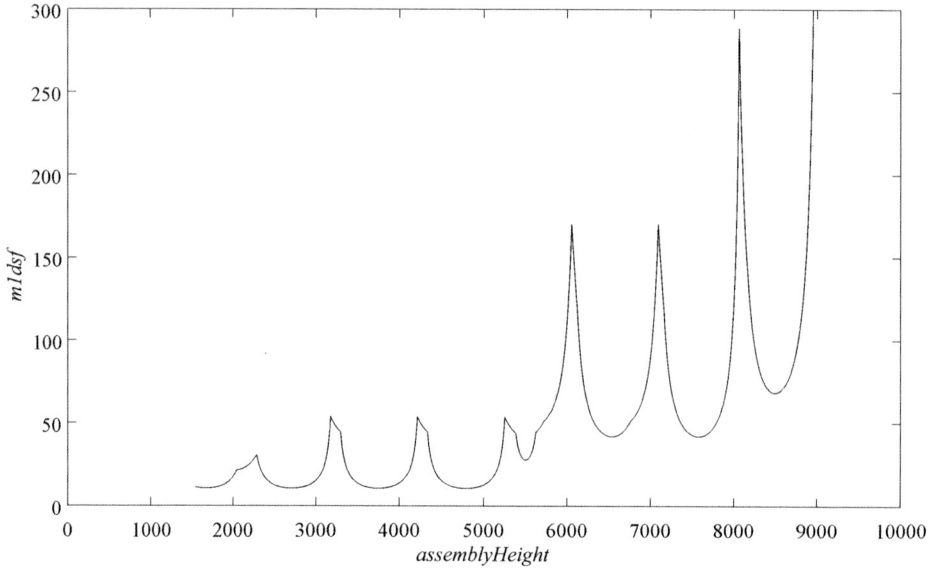

FIGURE 3: Distortion scale factor as a function of assembly height.

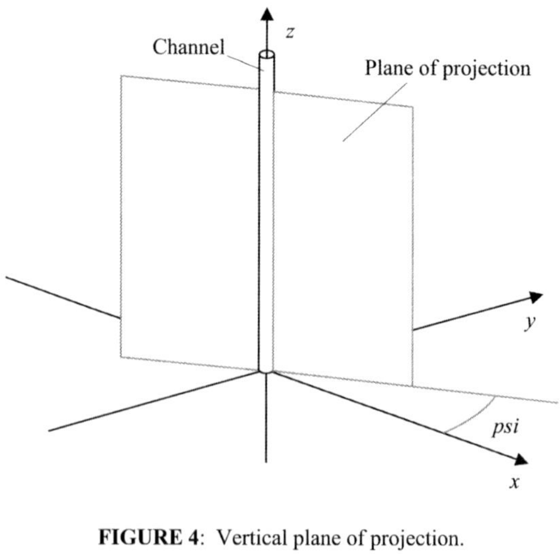

FIGURE 4: Vertical plane of projection.

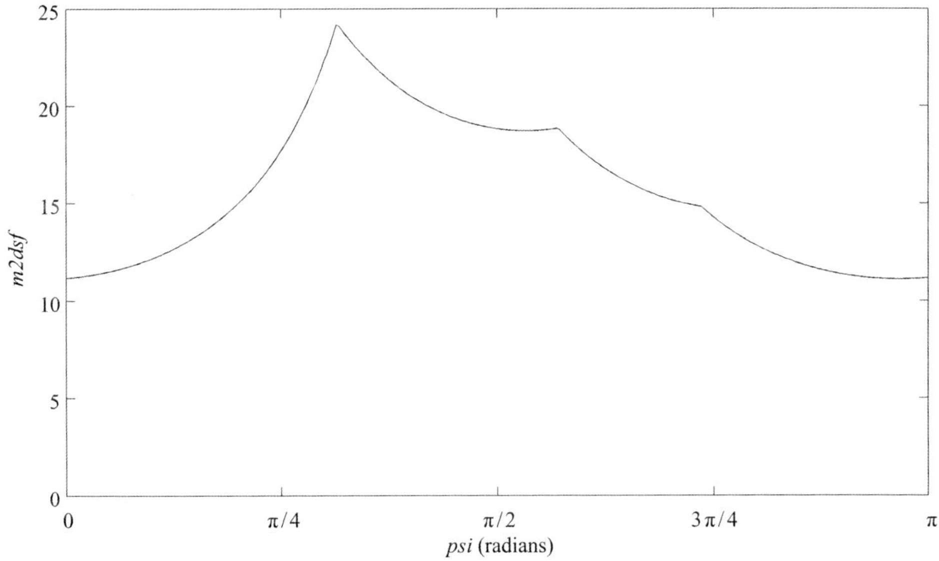

FIGURE 5: Distortion scale factor as a function of assembly height.

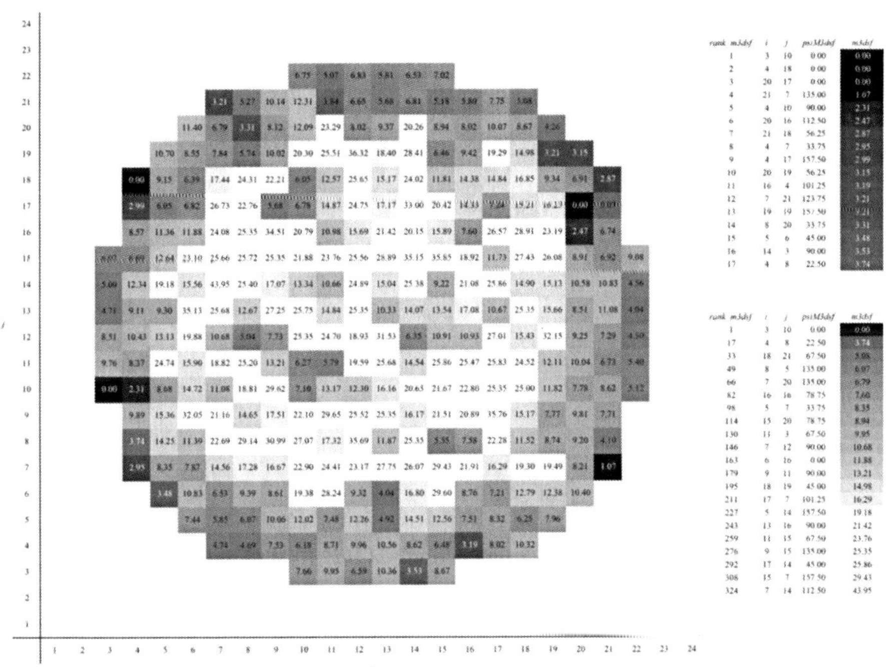

FIGURE 6: Core map giving *m3dsf* values for a 30% doubly cracked HRAHYA core with 5% axial-key failure.

SHORTCOMINGS AND CONSERVATISMS

LEWIS uses simplified geometry. In the reactor, the bore of the brick can vary with height. In its current form, LEWIS is not able to model this. In practice, the minimum radius is taken making the assessment conservative. LEWIS does not model forces. In the reactor, the assembly will be able to bend slightly. By not modelling this, LEWIS is conservative.

CONCLUSIONS

The LEWIS method assesses the functionality of AGR-core channels by calculating the maximum factor by which the channel distortion can be scaled and still allow the passage of a control rod or fuel stringer. The relative speed of this method enables assessment of every channel in a whole-core model, which enables the production of core maps and the ranking of channels.

Acknowledgements

The author would like to thank EDF Energy for financial support and permission to present this paper.

References

Gass, S. I. (2003). *Linear Programming : Methods and Applications*, Dover (New York). ISBN: 048643284X.

Karloff, H. (1991). *Linear Programming*, Birkhauser (Boston). ISBN: 9780817648435

Press, W. H. (2007). *Numerical Recipes: The Art of Scientific Computing*, Cambridge University Press (Cambridge), ISBN: 0521880688

Shaw, D. J. *et al.* (2008). Application of whole core modelling methodology with representative core loading to life extension of AGR graphite. In: Neighbour G. B. (ed.) *Securing the Safe Performance of Graphite Reactor Cores*, RSC Publishing (Cambridge). pp. 23–28 ISBN: 978-1-84755-913-5.

Part E

Open Discussion

Open Discussion

Anthony J. Wickham
The University of Manchester, Oxford Road, UK. M13 9PL
Email: confer@globalnet.co.uk

Abstract
This paper records the general discussion that followed each topic session. Each session was facilitated by moderators who initially opened the discussion and then aided the discussion by seeking opinions and / or comments to establish the current "state of the art" of the topic.

Keywords
Discussion, state-of-art, topic session

MECHANISTIC

Facilitators: Professors John Knott and Paul Bowen, University of Birmingham
Professor Rand's invited lecture discussed the main aspects requiring mechanistic understanding. These being:

- Structural basis for dimensional change because it controls internal stress in components.
- Irradiation creep as it alleviates internal and externally imposed stress.
- Structural basis of properties.
- Radiolytic oxidation and its effect on structure.

In his lecture, he reported that dimensional change and property changes are a consequence of structural modifications induced by the combination of irradiation damage and radiolytic oxidation and what are measured mostly are physical property changes and density. He stated that there is limited understanding of the structural changes bringing about these effects, especially at medium to high fluence (*i.e.* in the region of dimensional change turnaround and beyond). Consequently, there was an emphasis upon empirical modelling procedures and extrapolation, and so confidence to predict properties is limited. The reason for the lack of mechanistic understanding, at least partly, is the structural complexity and inhomogeneity of polycrystalline graphite. As properties are macroscopic, and Professor Rand's lecture stated the need for macroscopic structural characterization, it was important to understand the position where collectively we have:

- A fair amount of nano-structural characterisation.
- Very limited micro-structure characterisation.
- Limited macro-structure characterisation (porosity).

Consequently, there was a need to understand how macrostructure is determined by micro- and nano-structural changes and this resulted in the following key questions:

- Can we define a structural repeat unit at the macroscopic scale that can characterise the essential features controlling the properties of graphite?
- Can we determine the structure-property relationships at the nano- and micro-scales to predict the macroscopic properties of the characteristic unit?

- Are multi-scale modelling procedures well enough developed to enable this prediction?
- If not, are they likely to be in the foreseeable future?

Professor Knott referred to the questions which had been raised in Professor Rand's lecture, commencing with the essential question 'to what extent can irradiation damage be regarded as the "reverse" of the graphitisation process?' He commented on the sp^2 nature of the 'ruck and tuck' bonding and contrasted this with the sp^3 nature of the spiro bridges.

Professor Heggie commented that the sp^3 spiro bonding made the pinning process almost certain, but considered that, above 200°C, slip would also be enabled. The higher temperature effects in Young's modulus either had to come from rapid irradiation-damage processes, which was perhaps difficult to believe, or from a collective property of dislocations, as in work hardening.

Professor Rand added that he could see the structure 'breaking up' under irradiation and becoming almost turbo-stratic: this must surely alter the possibility for basal-plane slip to occur. There is a decrease in modulus as you go to the graphitised state. He asked whether the effect of irradiation could be considered as the reverse of the graphitization process, and what might be learned about irradiation damage from the synthesis processes.

Dr Bradford said that he was not aware of any low dose information on thermal conductivity: the processes include scattering at grain boundaries. It would be interesting to see how quickly a single crystal degrades. If it is rapid, the driving mechanism should be within the crystal; if slow, it would be due to effects between crystals.

Professor Knott commented that changes in coefficient of thermal expansion were essentially the stiffness of bonds responding to the changes in temperature.

Dr Bradford said that more work needed to be done to investigate the balance between single-crystal and polycrystalline behaviour. This led Professor Rand to wonder how one should define a crystallite. Two different definitions existed in the literature and relate to the X-ray coherence length: HOPG is made up of coherent regions, not one single crystal. Hence, HOPG really models a *domain* displaying a preferred orientation over hundreds of micro-metres. Professor Flewitt asked if this had any effect on dislocation mechanisms since the crystallite sizes are tens of micro-metres and their mobility may be modified. Professor Heggie suggested that 'basal slip' might be a better way of describing the process, rather than dislocations. Professor Rand commented that as we see the structure becoming more turbo-stratic this would lead to more twisting of layers and possibly to changes in Coefficient of Thermal Expansion. Dr Bradford asked for a definition of 'turbo-stratic', prompting Professor Heggie to offer the suggestion that it is 'the least disordered one can get', when stacking coherence between one plane and another is first lost – each layer having random orientation about the *c*-axis.

Professor Marsden commented that we seem now to be using the term 'turbo-stratic' to describe irradiated graphite: was it sure that we were really applying the term correctly? Professor Knott remarked that Professor Rand had also questioned the description of the effects of irradiation on filler particles, wondering what is the relative 'stiffness' of the matrix and filler particles. Professor Rand proposed that, in arriving at a model, one should consider the matrix and the filler particles separately.

Dr Haag said that the separate irradiations of matrix and filler material had been tried: it was easier to manipulate binder coke samples because one had needed to 'glue' filler particles together. As a result, tests had been conducted using different percentages of binder. The more binder which was present, the more the sample shrank under irradiation. A difference was seen in the degree of graphitisation even if samples had all seen the same (graphitisation) temperature: Professor Heggie asked whether this could also reflect a different degree of porosity. Dr Haag said that filler coke was usually selected to get the best graphitic structures after the graphitisation stage of the production process whereas the requirement for binder material is to make the best 'glue'. The binders used for subsequent impregnations had to be more liquid in order to facilitate penetration: consequently, there were at least three different levels or (or definitions of) 'perfection' in graphitisation. These potentially different phases could not be investigated by X-ray diffraction because X-rays only detect the graphitised parts. Professor Flewitt pointed out that micro-scale mechanical tests at the University of Bristol on irradiated filler particles in PGA graphite supported a higher elastic modulus.

Dr Roach suggested that new experiments should perhaps be conducted separately as part of the FUNGRAPHITE project, and outlined some possible instrumental techniques which could be applied to study the materials.

Mr Steer was concerned about discussions of single-crystal behaviour when we were actually concerned with 'billions of them'. He felt that there must be some 'half-way house' which could be considered: a crystallite surrounded by chemical bonds perhaps. It would be interesting to understand the nature of this bonding... in other words, how the 'glue' works. Professor Heggie said that planned work would try to look at that through first-principles studies of graphite grain boundaries. Dr Knowles noted the appearance of cracking after turnaround in the filler particles: clues may lie there. Professor Knott noted that this should also take account of the 'bubble-like features' noted by Dr Haag as developing around turnaround, suggesting that they arose from the amalgamation of smaller 'bubbles'. Professor Heggie had asked how they might possibly move around, and Professor Knott referred to such a possibility in the thermal-spike region. Professor Heggie responded that the thermal spike is not thought to be important for fission neutrons in graphite.

Dr Bradford wondered how low crystal stress can have a large effect on strain rate, and how such a modification should be built into models. Professor Knott commented upon Dr Metcalfe's recent results which helped to resolve long-standing questions about factors in the description of thermal expansion. Professor Marsden referred to forthcoming experiments to low irradiations in which HOPG would be included.

Prior to Professor Knott concluding the discussion, Professor Flewitt pointed out that a key mechanistic challenge identified by Professor Rand remained, namely to understand the difference between stress controlled dimensional change and irradiation creep since it has significant plant implications.

EMPIRICAL

Facilitator: Dr. David Knowles, Atkins

Dr Knowles suggested that given the subject areas covered in the presentations, the discussion should focus on three key areas. Firstly, noting Professor Marrow's lecture, and the noting of the influence, for example, of bi-axial stress states in crack growth; do we have enough information to form the basis of brick-failure predictions? Further, he posed the question as to whether we could ascertain if keyway-root cracking would proceed in a rapid manner or whether it would initiate with an element of sub-critical crack growth. A seismic event would probably result in initiating a very rapid crack 'at power' or off-load. A further point he raised by the question as to whether sufficient quality and quantity of irradiated / oxidised data existed and what more was required?

Professor Knott said that to start with it would be sensible to seek to explain the nature and generation of the cracks we have actually seen (particularly bore cracking) rather than the forms we have not yet experienced. He stated that we are aware of the degradation in modulus: thus the driving force for thermal strain is reduced, but we are still some steps away from a full understanding of the processes.

Dr Knowles asked if the Blackstone experiment was sufficient to fill the gap. Professor Knott thought not with more needing to be done on changes in strength under continued oxidation and irradiation. Professor Flewitt asked if we have enough information relating to size effects (small sample data to be applied to large components). Professor Knott argued that test specimens need to be designed be larger in size (although in the irradiation tests this is impractical – *Editor*).

Dr Knowles then commented that when one considers future predictions of behaviour at keyways, the available understanding being applied is very crude, based upon relatively simple notions of notch strength and its reduction with in-service exposure and local residual stress evolution. Professor Bowen commented that stress fields in large components are extremely complex and, for the keyway position, we do not really have 'virgin' data. He asked whether it was sensible to have a complex model when one does not really understand the behaviour in small samples. Optimistically, though, he felt more comfortable about weight-loss data: did not high weight loss simply mean more, bigger, pores. He urged the meeting to be careful in defining what one can and cannot do.

Dr Bradford said that we have heard a lot about graphite being quasi-brittle in Professor Marrow's lecture, but then later that irradiated material behaviour appeared to be elastic. He wondered if the universities should be moving their focus of work, with the most usefully to be investigated now subtly changed. Professor Marrow essentially agreed. He added that although it is plausible that quasi-brittle behaviour may return at high weight loss, data on the fracture behaviour, particularly at stress concentrations, are needed for this regime.

Professor Knott noted that dimensional-change results which had been presented by Dr Hall showed little sensitivity to (or evidence of) turn-around and hence whether significant tensile stresses would evolve at the keyway roots. Mr Steer cautioned that other parameters showed a much greater sensitivity to stress reversal.

Dr Moskovic agreed with Dr Bradford that the industry needed to tackle more irradiated / oxidised material in relation to quasi-brittle behaviour noting that the graphite was more and more tending to this type of behaviour. Professor Flewitt suggested that crack propagation could be thought of as a progressive 'ligament failure' which we needed to understand better through correlation of experimental observations and numerical modelling. Dr Bradford also noted that the behaviour of PGA and Gilsocarbon graphites at high weight loss appeared to be different, and that is surely telling us something, in that this is the region which we should be focussing our efforts on.

Mr Steer did not agree that Gilsocarbon material was becoming quasi-brittle: quasi-brittle material is extremely weak with a relatively high work of fracture, whereas graphite (like glass) has high strength, but a low work of fracture. The issue to be considered is not just the material property, but how a sample (or component) is loaded. A large brick component contains several tens of Joules of energy to be dissipated.

Mr Banahan said that EdF Energy (UK) has cracked bricks in the AGRs: could this cracked graphite material be sampled? Dr Reed confirmed that cracked bricks have been trepanned, but not at the position of the cracks. A trepanned area has been removed however where bore cracking was observed subsequently. Mr Steer said that the samples from cracked bricks did not show any obvious differences in mechanical behaviour and Dr Bradford concurred with this.

Professor Knott reflected further on the latent energy within the graphite: when you scale up a sample to component size, you have not actually scaled up the cracks *pro rata*. Dr Knowles noted another important difference: in the reactor whereby stresses are ramped over very long periods of time whereas in laboratory tests the time of load application is very short. Real components may 'store up' an internal energy for a while and then eventually crack or slow crack growth may occur releasing strain energy in a more controlled manner. Mr Steer noted that some bricks fail at one half of the expected loading. Returning to the notion of quasi-brittle behaviour, Mr Moskovic made a comparison of the load-displacement curves of deeply notched and un-notched material, suggesting that it became less brittle as you put in the notch.

Dr Knowles asked, in a perfect world, what work we should undertake now. Dr McLachlan reminded the meeting that most ideal solutions were, in reactor, extremely impractical. Mr Lamb asked whether work on stress-concentrating features was being sponsored at appropriate universities. Dr Poulter commented that, despite the discussion highlighting what was *not* understood about cracking, the rate of bore cracking appeared to be relatively well-behaved, not being observed to increase rapidly between outages; this was why the ONR was allowing AGRs to continue in operation despite the bore cracking. Indeed, regulatory concern was more about keyway root cracking given that, post-turnaround, the stress is tensile on the outside of the bricks. He then asked the question "given that we do not understand the contribution of stress-concentrating features well, could there be any situation where the rate of cracking would become much faster such that one could be 'caught out' between outages?" In summary, he was worried about the uncertainties in the *rate* of cracking.

Dr Knowles proposed a further topic. Relating to the back-end process of modelling ('tuning' by back-fitting against plant data), it would be interesting to see more evidence that this was statistically supported. His observation was that actual validation data was very limited. Even looking at data forthcoming from the Blackstone experiment, he felt that the evidence would be sparse. Mr Moskovic commented that only Mr Chidwick's paper had involved an analysis of extensive data. He wished to try and identify the variables that required better definition. Mr. Robinson felt that the data were *not* sparse. Mr Chidwick said that, in fitting models to the data we have, one needed to address alternative model.

Dr Knowles proposed that irradiation creep should be thought of as stress-dependent dimensional change. Looking again at Dr Hall's work this was an important variable within the dimensional change which would need to be incorporated to clarify its influences on behaviour. Further, a range of forms of relationships encompassing multi-axial stress state influence may be considered. As such using an emulator could be seen as a powerful tool allowing focus on the problem from both ends.

Dr Reed suggested that what Mr Chidwick was concerned about in his work was not what happens, but rather not knowing what *might* happen. This led to the thought: 'what would have to happen in order for the unthinkable plant state to arise?' Again, this was a powerful application of the emulators, which should be the future approach.

Dr Bradford was struck by the hydrostatic compressive deformation that graphite could tolerate and still recover based on one of the talks at the conference. Professor Knott pointed out that this was often observed in materials under compression and did not surprise him.

STATISTICAL

Facilitator: Dr Martin Lamb, ONR

Dr Lamb suggested that the discussion should deal with the treatment of the major AGR safety issue (from the perspective of the ONR) which is keyway-root cracking, which is only tolerable because of the expectation of a progressive onset. He suggested that dependence on deterministic models was probably unrealistic and hence appealing to probabilistic predictive methods. Another side the question was the acceptance of both single and doubly cracked bricks: he noted the importance of diversity of models to predict the acceptability of damage and referred to the SOLFEC methodology as presented by The University of Glasgow in this context. ONR was looking for the best means to quantify uncertainty and the scope for the validation of predictions.

Dr Robinson referred to the distribution of irradiation dose around the cores, indicating that we will indeed appeal to variability, but must be careful not to include parameters which are not actually variable, as this would lead to misunderstanding in the conclusions drawn. He wondered if there were specific instances where such misunderstandings had been sought and examined critically.

Dr Crawford wondered how much of the issue was to do with variability when testing the graphite. Does the instrumentation employed introduce even more uncertainty to that inherently present in the material? Professor Flewitt noted that it was important to understand the origins of uncertainties and therefore one needs knowledge of the underlying measurements in order to appreciate the evidence fully. Dr Maul pointed out that it is always very clear from the statistical methods if you have got the variability wrong, and that there had been four iterations of the protocols already in the course of the Quintessa development work. Mistaking uncertainty for variability could be non-conservative.

Professor Marsden said that we are seeking to define the incidence of keyway-root cracking without actually sampling or measuring anything, and there were also large uncertainties in weight loss in a brick, so one needed to know something about *every* brick. Dr Robinson asked if component shape was relevant. Dr Reed said that EdF Energy (UK) would struggle to follow Professor Marsden's approach, and Dr Poulter agreed that it was obviously impractical to inspect everything at every outage: he wondered if it was better to inspect the same channels repeatedly in successive outages or whether it was better to go to completely new ones. Drs Maul and Robinson pointed out that if you can define precisely what information you want, the statistical tools will define what the better inspection strategy is.

Mr Steer commented on variability, noting the importance of understanding any internal correlations between variables which could contribute to the result. Dr Reed referred to the unexpected 'dose squared' dependence arising in Dr Moskovic's work and Professor Wickham's challenge to that on the basis that the scientific understanding made it totally illogical: Dr Reed also noted the result from the Health and Safety Laboratory suggesting that bricks in some channels were more likely to crack than others: neither outturn would have been foreseen, indicating that one might not know the appropriate variables on which to focus without the benefit of some experience. It was all about 'knowing where to look' for the essential evidence.

Mr Steer suggested that the tails of distributions were perhaps more significant than the means, yet we tend to focus on those mean values. Mr Moskovic concurred, and spoke about the number and nature of measurements needed to define a distribution. Dr Lamb agreed that one should not be misled by 'extreme' data points but, at the same time, misplaced complacency was dangerous: we could be in a position of feeling comfortable about a process and then something came along between outages which strongly challenged that position. Dr Maul stressed the importance of feedback between monitoring and statistical modelling and said that statistical tests could be used to identify the onset of damage processes.

Returning to the subject of distribution functions, Mr Steer considered that when mechanisms were not understood, one could be 'led astray' by the distribution tails perhaps influenced by the presence of weak material: deterministic methodologies can lead to the wrong conclusions. The initial data points emerging for some property may not be part of the main population (*e.g.* weaker components failing first). Dr Novovic drew attention to analogies and differences between bore cracking and keyway-root cracking.

Mr Steer spoke about explaining the cracking we *didn't* get. Such issues raise doubts, and one should do more things to resolve such doubts. Considerations of statistical treatments are throwing up doubts about deterministic methods. Dr Reed urged that this should be part of a full feedback loop, and Dr Bradford concurred, commenting that the statistical approach aids the deterministic one and *vice versa.*

Dr Lamb introduced irradiation creep into the discussion. Creep is a strong predictor of stress and might affect both variability and uncertainty: were these being confused, given the sparsity of data? Dr Novovic commented upon the public's general inability to interpret probabilistic arguments appropriately.

Mr Banahan referred back to the significance of best-estimate data *versus* the probabilistic approach: Mr Slater remarked that the current work on keyway-root cracking is based upon best estimates.

During the preceding technical session, Professor Wickham had speculated on whether it was appropriate to wring everything possible out of the statistical treatments before considering the impact of individual pieces of data from the technical point of view, or whether the two processes should proceed independently and in parallel. Mr Maul, in the opening of his presentation, had himself posed the question for what he described as 'high-risk' situations, whether there is a point at which you must turn away from statistical treatments in favour of deterministic solutions.

Dr Lamb said that he was absolutely confident of the need to be cautious and to distinguish 'uncertainty' and 'variability': one was on safer ground when looking at field variables (effects of temperature, dose, *etc*). There was also no doubt that the statistical approaches should be further developed, and that the debate had been a healthy reminder of the need to 'bring a rigour to bear'. Mr Steer said that it was useful to be more sceptical and this could lead to a greater degree of acceptance of the outcome. There was also a clear need for improved mechanistic understanding and good quality materials data, particularly for irradiation creep.

PLANT

Facilitator: Dr Gareth Neighbour, Oxford Brookes University

Dr Neighbour stated that Mr Banahan's lecture had 'brought some reality' to the situation, and brought to life the challenges which underwrite reactor operation. With now quite lengthy periods of operation, hindsight revealed that the most appropriate courses of action had not always been followed. For example, he referred to the 'trade-off' in coolant composition in Magnox plant between the competing needs for the inhibition of graphite oxidation and steel oxidation. This experience was enabling a more coherent approach to present decision making, reviewing underlying assumptions in the light of new knowledge and experience, and striving to achieve simple technical arguments out of complex information. The presentation from Strathclyde University illustrated how it was possible to extract the maximum benefit from monitoring and operational data.

Mr Poulter commented on the 60 years of graphite-reactor experience and the regulatory need for 'defence in depth'. He posed the question 'do we know as much as we think we do?' and warned against the threat of 'Groupthink', in which one's thinking can tend to fall into line with others because the brain tends to make assumptions that the 'others' must have reached their course of action after a full analysis, which may not be the case. He also asked the meeting whether it was possible that models were now so complex that validation was impossible. He also wanted to recognise the richness and inventiveness of the material presented, and stressed the value of the meeting in obtaining peer review and in giving greater confidence to the regulators.

Mr Banahan had posed a number of issues in his presentation, and was asked whether he felt that the subsequent presentations had resolved them. He felt that the answer was 'not entirely': clearly there were things which still needed to be done, but was unsure (in the Magnox Ltd. context) whether the stations were the right people to address them, *e.g.* to use the developed software to make best use of monitoring data. In the AGR context, Dr Reed considered that EdF Energy had not found such problems yet, because current core 'damage' was tolerable and so the system was not yet 'stretched'.

Mr Poulter agreed that some of the questions which he had had in mind had also been answered in the subsequent presentations. He gave the example of the use of fuel-stringer load-trace data, regarding it as a neat way of accessing information which the reactor itself could provide. Dr Reed said that, in principle, as a result of extensive inspection obligations, Magnox now had a 'fingerprint' of every channel in the remaining reactors, and now was the time to look for any changes in those channels whenever opportunities presented themselves.

Dr Bradford referred to the potential loss of historical information for reasons such as fading of ink on old chart recordings. Archiving of potentially useful information in non-degradable formats was essential. Returning to the point about trading off the relative merits of modelling and measuring, one should remember that what is actually being traded is uncertainty. Not only is the analysis of graphite behaviour very complex, but also is the way in which we seek to use such information.

Dr Neighbour asked whether we really understand how each model fits into the overall scheme of things. In response, the discussion raised the concern that some examples seemed to present complex detail just for details' sake. Models are only 'hooks', or aids, upon which to distribute ones understanding: they do not have to be perfect all the time. In countering this point, Mr Steer observed that one sometimes notices different trends which turn out to be highly significant. He thought that Dr Bradford had been very successful in pulling apart complex data, and looking at it from more than one angle.

Dr Brighton observed that there is clearly some information which could be regarded as 'rock solid' and other information and models which offer support. He did not like to see 'complexity heaped upon complexity', seeking to have clear safety cases built upon transparent presentation rather than upon complex arguments which were difficult to unravel.

In response to a question about the role of the academic representatives, Professor Marrow suggested that their role should be to look a little further ahead than the companies, and try to stay one step (at least) ahead of the game. Dr West agreed, and suggested that there were more data yet which could be tapped into in the way described in his paper. It would also be useful to marshall diverse sources of data together, because judgements made upon their basis would be likely to be sounder. Professor Flewitt summarised this position: monitoring data are invaluable, sound modelling is good, and academic work supports future predictions.

Dr Neighbour commented upon the different levels at which information as acquired and was needed: from the individual sample through the reactor as a whole. One needed to be careful to see the whole picture. Mr Mike Davies reminded the meeting that, at the design stage, everything had been a compromise because of lack of experience and data. He remarked that, had it been possible to get everything right at the design stage, then the present meeting would hardly have been necessary. As an example, he cited the significant change in coolant composition which had been found necessary compared with the assumption at the design stage. This impacted strongly on graphite oxidation behaviour: in addition, the impact of thermal stresses had not been fully recognised, and so forth. Feedback loops played a large role now in ensuring the continuance of safe reactor operation. Dr West agreed that, had predictions of behaviour been available from the start, it would have been much better than launching into investigations half-way through reactor operating life. Dr Neighbour wondered whether we were currently making the appropriate level of predictions: did we need to do more (or possibly, less)? Mr. Robinson commented that, at whole-plant level, there was not much that we could do. Mr Maul pondered on the nature of the measure of 'whole-core performance'. He felt that the operators were still struggling to find that, although it was something that was clearly needed.

The meeting reflected briefly on whether it might be possible to learn from overseas operators, but there is little relevant information or experience outside the UK now. Dr Reed offered an example of 'groupthink failure' based upon experience in Russia. Mr Booler thought that we should be looking closer to home. To aid the AGRs, look hard at Magnox. Both Mr Banahan and Dr Reed said that this already happened, with Magnox representation on EdF Energy's Graphite Core Committee given as an example. The question of whether EdF should sponsor a major trepanning campaign in a shutdown Magnox reactor was also raised. This could be an exceedingly expensive activity and the value and necessity would need careful thought through. Dr Poulter said that, should this be a desirable idea, the time for action was clearly the present. Mr Steer doubted its value given the disparity in so many contributing factors, such as graphite type, temperature, irradiation flux, component design and loading, *etc*.

Author Index

Subject Index (by Keywords)